Frontiers of Environment

Frontiers of Environment

Issues in Medieval and Early Modern India

Edited by
Meena Bhargava

Orient BlackSwan

FRONTIERS OF ENVIRONMENT: ISSUES IN MEDIEVAL AND EARLY MODERN INDIA

ORIENT BLACKSWAN PRIVATE LIMITED

Registered Office
3-6-752 Himayatnagar, Hyderabad 500 029 (Telangana), INDIA
e-mail: centraloffice@orientblackswan.com

Other Offices
Bengaluru, Bhopal, Chennai, Guwahati, Hyderabad, Jaipur, Kolkata, Lucknow, Mumbai, New Delhi, Noida, Patna, Vijayawada

ISBN 978 93 86392 78 7

025412
11042017

Typeset in
Bembo Std 10.5/12.5
by Manmohan Kumar, Delhi

Printed in India at
Glorious Printers, Delhi

Published by
Orient Blackswan Private Limited
3-6-752 Himayatnagar, Hyderabad 500 029, Telangana, Indiae-mail: info@orientblackswan.com

Contents

Tables

Publisher's Acknowledgements

The essays in this volume have been edited, reworked and updated, and in some cases have had their titles modified by the respective authors. For permission to reproduce copyright material in this book, the volume editor and publishers wish to make the following acknowledgements:

SALIM A. ALI, 'The Moghul Emperors of India as Naturalists and Sportsmen'. Originally published in *Journal of the Bombay Natural History Society*, vol. xxxi, no. 4, 1927, Part I, pp. 833–45.

M. A. ALVI and A. RAHMAN, 'Introduction'. Originally published in M. A. Alvi and A. Rahman, *Jahangir – The Naturalist*, New Delhi: National Institute of Sciences of India, 1968, pp. 4–14.

CHETAN SINGH, 'Forests, Pastoralists and Agrarian Society in Mughal India'. Originally published in David Arnold and Ramachandra Guha, eds, *Nature, Culture, Imperialism: Essays on the Environmental History of South Asia*, New Delhi: Oxford University Press, 1995, pp. 21–48.

ARCHANA PRASAD, 'The Baiga: Survival Strategies and Local Economy in the Central Provinces'. Originally published in *Studies in History*, vol. 14, no. 2, July–December 1998, pp. 325–48.

BHANGYA BHUKYA, 'The Twilight World of the Caravan: The Regulated Market Economy and the Caravanners'. Originally published in Bhangya Bhukya, *Subjugated Nomads: The Lambadas under the Rule of the Nizam*, Hyderabad: Orient BlackSwan, 2010, pp. 27–63.

MAYANK KUMAR, 'Situating the Environment: Settlement, Irrigation and Agriculture in Pre-colonial Rajasthan'. Originally published in *Studies in History*, vol. 24, no. 2, July–December 2008, pp. 211–33.

TRIPTA WAHI, 'Irrigation, State and Society in Pre-Colonial India'. Originally published as Occasional Paper, *History and Society*, New Series 36, New Delhi: Nehru Memorial Museum and Library (NMML), 2013, pp. 1–42.

MEENA BHARGAVA, 'Changing River Courses in North India: Calamities, Bounties, Strategies—Sixteenth to Early Nineteenth

Centuries'. Originally published in *The Medieval History Journal*, Special Issue: Coping with Natural Disasters in Pre-Industrial Societies, vol. 10, nos 1 and 2, January–December 2007, pp. 183–208.

SIMON DIGBY, 'The Elephant in Medieval Indian Warfare'. Originally published in Simon Digby, *War-Horse and Elephant in the Delhi Sultanate, A Study of Military Supplies*, Oxford: Orient Monographs, 1971, pp. 50–82.

ASOK KUMAR DAS, 'The Imperial *Cheetahs* in Akbar's *Shikarkhana*, Seen through the Paintings of his Time'. Originally published in *The India Magazine*, vol. 5, no. 9, August 1985, pp. 28–39.

MAHESH RANGARAJAN, 'The Hunt and the Wilderness in Mughal India'. Originally published in Mahesh Rangarajan, *India's Wildlife History: An Introduction*, New Delhi: Permanent Black, 2005, pp. 11–21.

Introduction

Meena Bhargava

The complexity and diversity of environmental history, intense and contentious debates on themes and issues, related nonetheless, primarily to modern and contemporary times led to the introspection and a venture into understanding environmental issues in the medieval–early modern period of Indian history. This became imperative to resurrect the significance of this phase of history for a more viable assessment of the environmental processes. In the context of environmental history, till at least a few years back, the pre-colonial period of Indian history was spoken of disparagingly, as though devoid of history and considered a mere standard or a reference point to comprehend the developments, changes or continuities in the nineteenth and subsequent centuries. However, the task of creating a reader such as this one was a difficult one, considering the vast and colossal research, interpretations and approaches to environmental history. Given the enormity of the discipline, the Introduction focuses on aspects it considers valuable and worthwhile for the appraisal and reappraisal of environmental issues across time.

There are four parts to the Introduction. Part I argues against the notion of a 'static' pre-colonial period and demonstrates the redundancy of the bias against it in evaluating environmental issues, followed by a discussion on different trends in environmental historiography in Part II to facilitate and sharpen the assaying and interpretation. The environment has often been studied in segments focusing on one idea or feature at a time such as forests, wildlife, bio-diversity or simply the environment. It is peremptory to break the rigid binary categories to perceive the environment and provide an integrated way of thinking about environmental issues encompassing perceptions that surround the vital nature–wildlife–human interrelationships. With this in view, Part III of the Introduction discusses the relevance of the concept of 'agrarian environment' on the basis of the writings that establish pervasive links between the agrarian and the environment and insist that the exclusive treatment of the environment and the agrarian, each independent of the other, is to fail to understand either. Part IV carries a brief review of essays selected for the volume, as samples that explain the nuances of the medieval–early modern environment, unfold the specificities and generalities of different regions and provide comparative analysis therein on nature, forests, pastoralism, tribes, irrigation, water and wildlife.

I

It may be plausible to argue that the elucidation of environmental issues can be made less prejudiced with the dismantling of the customary practice of conforming to periodisation and retrieving the lost vocabulary of the pre-colonial period, in this case the medieval–early modern period, not merely as an allusion for the evaluation of colonial and post-colonial issues but to estimate it in its own right and perspective. Divisions and compartments between natural and humanistic disciplines and more notably labels such as ancient, medieval, modern, contemporary, or the past being viewed by many scholars as *tabula rasa*, prevent the connections that are significant and necessary to be established. The objective should be not merely to historicise environment but to also contextualise it in *longue duree* and emphasise on a wider vista, a dialogue, perhaps a synthesis and a smoother understanding of developments and transition of knowledge to appreciate a millennia-long history of India's ecosystems and their interface with human desires and ambitions, triumphs and failures.[1]

The Orientalists had stressed upon the 'dynamism' of the West versus the 'stagnant' East to underscore the bounties of the colonial rule and promote the belief that the overall structure of pre-colonial India had remained unaltered. Contrary to this, however, constant variations and transformations have been noticed within the ecological

[1] Mahesh Rangarajan and K. Sivaramakrishnan, eds, *India's Environmental History, From Ancient Times to the Colonial Period*, vol. I, Ranikhet: Permanent Black, 2012, p. 4; also see their (eds), *Shifting Ground, People, Animals, and Mobility in India's Environmental History*, New Delhi: Oxford University Press, 2014, pp. 7, 13–14; also Mahesh Rangarajan, *Nature and Nation: Essays on Environmental History*, Ranikhet: Permanent Black, 2015.

perimeters of India's limits and inner frontiers. Most pre-colonial polities emerged on the roads and crossroads of commercial and nomadic interaction.[2] The pre-industrial or pre-colonial past may not have been as idyllic, harmonious or benevolent as some have argued[3], but its ecological systems were transient, uncertain and constantly changing although all transmutations may not have been contributed by human beings; some modifications in nature could have been sudden and cataclysmic[4]. A variety of negotiations with nature, whether through technology or through the process of production, reflect the complexity and intricacies of human intervention in the pre-colonial environment and challenge the idea of homogeneity[5]. Regions of urbanism, ethnicity, empire, literature and territoriality made the land look very different in *c.* 1800 CE than in *c.* 1200 CE. There was a qualitative change in the way people thought about the land; for example, the history of the Tamil term for forest, *kadu*, indicates something about cultural change in agrarian societies in the Indian peninsula[6]. As opposed to the indexing like a 'pristine state of nature'[7],

[2] Jos Gommans, *Mughal Warfare, Indian Frontiers and High Roads to Empire, 1500–1700*, Warfare and History Series, London and New York: Routledge, 2002, p. 201; also Rangarajan and Sivaramakrishnan, eds, *Shifting Ground*, p. 10.

[3] See Madhav Gadgil and Ramachandra Guha, *This Fissured Land: An Ecological History of India*, New Delhi: Oxford University Press, 1992 and other such works that have been mentioned in the course of this Introduction. The works of Gadgil and Guha have been observed by many as polemical narratives. However, it is necessary to acknowledge their pioneering contribution to ecological history and the first coherent model of how natural resources were conserved and organised in pre-colonial India.

[4] Rangarajan and Sivaramakrishnan, eds, *India's Environmental History*, vol. I, pp. 1, 8.

[5] In the context of medieval and early modern Rajasthan, see Mayank Kumar, 'Ecology, Social Stratification and Agrarian Production in Medieval Rajasthan', in B. L. Bhadani, ed., *Medieval India 3, Researches in the History of India*, New Delhi: Manohar, 2012, pp. 257–270, especially p. 258.

[6] For more details, see David Ludden, *An Agrarian History of South Asia*, New Cambridge History of India, Cambridge: Cambridge University Press, 1999, pp. 148–150.

[7] Richard Grove observes that some micro-studies, mainly in Africa, have shown that the notions of the golden age of pre-colonial 'common rights' and 'common property resources' are largely mythical. See his *Ecology, Climate and*

the environmental and cultural developments in the Western Ghats during the fifteenth to the seventeenth centuries, for instance, were far from static. It witnessed changing ecological contexts with forest loss and regeneration, alterations in the composition and distribution of vegetation and soil movement and also oscillating long-distance and local exchange involving forest products[8]. The boundless diversities in environmental situations like types of soil, patterns of rainfall, floral and faunal conditions preclude any fixed archetype on an all-India basis even if the agricultural development in early India may often appear repetitive. In other words, with extreme variations in climates, ecology and environment, the possibility of plurality can hardly be lost sight of in the Indian sub-continent.[9] In addition, the evolution, growth and sustenance of the human settlements and the human-nature interaction in different ecological contexts beyond the categories of region, regional polities and temporal divisions can be suitably illustrated by the study of monsoon ecology and its erratic nature.[10]

The environment has facilitated the understanding of not only demography and settlement patterns but also the ecological change since *c.* 1600 CE that can be grasped by computing the extent of

---

Empire, The Indian Legacy in Global Environmental History, 1400–1940, New Delhi: Oxford University Press, 1998, p. 186.

[8] Kathleen D. Morrison, 'Environmental History, the Spice Trade, and the State in South India', in Rangarajan and Sivaramakrishnan, eds, *India's Environmental History*, vol. I, pp. 296–326.; also see her article, 'Conceiving Ecology and Stopping the Clock: Narratives of Balance, Loss and Degradation', in Rangarajan and Sivaramakrishnan, eds, *Shifting Ground,* pp. 39–64 where she provides evidence from the Ganga and Indus river basins and also the Vijayanagara region to counter the notion of the primeval, pristine and untouched forest.

[9] Ranabir Chakravarti, 'The Creation and Expansion of Settlements and Management of Hydraulic Resources in Ancient India', in Richard H. Grove, Vinita Damodaran and Satpal Sangwan, eds, *Nature & the Orient: The Environment History of South and Southeast Asia,* New Delhi: Oxford University Press, 1998, pp. 88, 102.

[10] Mayank Kumar applies the 'larger umbrella of monsoon ecology' to study human–nature interactions and the peculiar identity of early modern Rajasthan and its people. See Mayank Kumar, *Monsoon Ecologies, Irrigation, Agriculture and Settlement Patterns in Rajasthan during the Pre-Colonial Period,* New Delhi: Manohar, 2013, pp. 15–16.

cultivation around that time and comparing it with figures of more recent times.[11] Medieval environmental history, argues Ellen Arnold, is a dynamic and growing field; while it may be rooted in 'older questions' of medieval history, it takes advantage of new and exciting sources and methods and explores themes such as climate and ecological history, the history of agriculture and water, landscape, and religious studies.[12] The so-called 'traditional pre-modern' societies or medieval ecologies were never static but marked by human adaptations to change. Small, often unnoticed innovations and changes gradually improved human life and productivity.[13] There were some long-established and fundamental relationships between society and its physical surroundings. No society could survive for any length without them, making imperative the understanding of the socio-economic processes and the specific ecological environment within which these relationships operated.[14] It may thus be appropriate to identify, describe, and reflect on the processes by which human beings intervened in the natural environment during the early modern period, that is the late fifteenth to the early nineteenth century, when the pace and magnitude of change increased in human societies in every part of the world. The term early modern suggests, as John Richards has argued, shared historical processes and long-term currents that accelerated in this period and

[11] Shireen Moosvi, 'Ecology, Population Distribution and Settlement Pattern in Mughal India', in Shireen Moosvi, *People, Taxation and Trade in Mughal India*, Collected Essays, New Delhi: Oxford University Press, 2008, pp. 89, 92–93.

[12] Ellen F. Arnold, 'An Introduction to Medieval Environmental History', *History Compass* 6: 3, 2008, pp. 898–916, in particular, p. 898.

[13] Richard Eaton, *The Rise of Islam and the Bengal Frontier, 1204–1760*, New Delhi: Oxford University Press, 1997; also John F. Richards, *The Unending Frontier: An Environmental History of the Early Modern World*, Berkeley: University of California Press, 2003, pp. 1–3; also see Chetan Singh, *Region and Empire: Punjab in the Seventeenth Century*, New Delhi: Oxford University Press, 1991 particularly for its significant contribution that suggests the importance of the integration of ecological and historical developments in the seventeenth century.

[14] For an understanding of the basic characteristics of society–environment interaction in the Himachal Himalaya, see Chetan Singh, *Natural Premises: Ecology and Peasant Life in the Western Himalaya, 1800–1950*, New Delhi: Oxford University Press, 1998.

had a deep impact on the rapidly moving anthropogenic change, or human-induced environmental change or the collective human action on the ecosystems of the world during the nineteenth and twentieth centuries. There are profound continuities that link the environmental history of the early modern and modern worlds.[15] Trends that began to gather the fundamental constituents of modern agrarian environments are sufficiently perceptible in the sixteenth century to justify using the phrase 'early modern' for the period *c.* 1550 CE–1850 CE.[16] Deploying money and horsepower, the Mughal Empire achieved unprecedented agricultural, commercial and political expansion. In terms of ideology and ritual too, the Mughal state reflected as much the 'world of the settled court as that of the nomadic war-band'.[17] Paul J. Crutzen, a geologist, considers the early modern phase as laying the foundation of 'the Anthropocene'—a new geological epoch characterised by intense human activity and the warmest climate in the history of the earth. Crutzen observes that this era began in 1784 CE when atmospheric carbon levels contained in ice cores first began to increase.[18] Within such constructs, the early modern period may easily be considered a period of global social, political and technological transitions in which human beings and the planet crossed into this *terra incognita* (unknown lands).[19]

II

The quest for tracing 'human ecological footprints'[20] entails the breaking of stiff, distinct categories for the understanding of the environment or environmental processes in India, 'reframe conservation, renew

[15] Richards, *The Unending Frontier,* pp. 1–3, 22.

[16] Ludden, *An Agrarian History of South Asia*, p. 122.

[17] Gommans, *Mughal Warfare*, p. 202.

[18] Paul J. Crutzen, 'Geology of Mankind', *Nature*, 415, 3 January 2002, p. 23.

[19] David Biggs, 'Review of John F. Richards, *The Unending Frontier: An Environmental History of the Early Modern World*, Berkeley: University of California Press, 2003, in *H-Environment, H-Net Reviews in the Humanities and Social Sciences,* May 2006, p. 1.

[20] See Rangarajan and Sivaramakrishnan, eds, *India's Environmental History*, vol. I, pp. 7, 24.

our connection with the earth, and reinvent the ecology that has sustained humans and wild nature alike'.[21] The emphasis should be on inter-disciplinarity and accessibility of environmental history and a unified, harmonious way of thinking about environmental issues. We perhaps need to ask—where does the *environment* begin and end? Any presupposition that there is a natural environment which is separate from the people who live in it and their culture creates an artificial barrier and the assumption that the concept of environment refers primarily or exclusively to the physical environment. But then, the physical environment is only one part of the environment which influences human beings and which they in turn strive to influence. There is also a social environment, in which humans are mutually constitutive of each other. Beyond this the interaction between the two environments is so tight that eventually the distinction between them disappears completely.[22] The environment, thus, cannot be understood as merely pretty trees and tigers, threatened plants and ecosystems—it is literally the entity on which man subsists, and on which his entire agricultural and industrial development depends. Human intervention may be successful in establishing a new and stable balance between society and nature, yet on other occasions, it may itself turn into the cause of ecological decline.[23] In other words, development without concern for the environment can be carried only to a point and cannot but be brief; in the long-term, it can only be anti-development and may result in increased human endurance and pain, poverty and oppression.[24] Environmentalism, therefore, needs to be viewed as a social programme, an agenda for action which seeks to protect habitats, protest against degradation and prescribe less destructive technologies and lifestyle.

[21] Ghazala Shahabuddin, *Conservation at the Crossroads, Science, Society and the Future of India's Wildlife*, Ranikhet: Permanent Black, 2010 in association with the New India Foundation, p. 206.

[22] Piers Vitebsky, 'A Farewell to Ancestors? Deforestation and the Changing Spiritual Environment of the Sora', in Grove et al., *Nature & the Orient*, p. 967.

[23] See Radhakamal Mukerjee, 'An Ecological Approach to Sociology', in Ramachandra Guha, ed., *Social Ecology*, New Delhi: Oxford University Press, 1994, pp. 120–31.

[24] Anil Agarwal, 'An Indian Environmentalist's Credo', in Guha, ed., *Social Ecology*, p. 347.

Environmental history, in other words, represents a multi-dimensional approach, involving the natural sciences, anthropology, cultural geography, sociology and history, facilitating a look at the catalysts for human change that have often been ignored. Defined as that which deals with the role and place of nature in human life, environmental history has more recently been described as 'an unevenly spreading blob'.[25] Formulations such as these reflect a wide array of methods and approaches, and studies from different regions and periods.[26] In fact, one of the significant features of environmental history is its competence to derive from the insights and techniques of several disciplines, and then to combine them in novel and often provocative ways of its own.[27] Environmental history is both a daunting prospect and a logical endeavour. It is daunting because trying to make sense of the entirety of history of human interaction with planet earth may take more than a lifetime as we sift through an ever-growing quantity of relevant information. It is logical because in the environment, as in history, everything is connected to everything else and nothing stands on its own.[28] To quote American environmentalist John Muir, 'When we try to pick out anything by itself, we find it hitched to everything else in the Universe'.[29]

Debates and investigations related to the environment and ecology[30] have led to a rapidly expanding discourse for over three decades. The

[25] Donald Worster, 'Doing Environmental History', in Donald Worster, ed., *The Ends of the Earth: Perspectives on Modern Environmental History*, Cambridge: Cambridge University Press, 1988, pp. 289–308; also H. Ritvo, 'Discipline and Indiscipline', *Environmental History*, Special Issue, 10: 1, 2005, pp. 75–76.

[26] To understand the multi-directions of environmental history, see Adam Rome, Michael Bess et al, 'Anniversary Forum: What's Next in Environmental History', *Environmental History*, Special Issue, 10: 1, 2005, pp. 30–109.

[27] David Arnold and Ramachandra Guha, 'Introduction', in David Arnold and Ramachandra Guha, eds, *Nature, Culture, Imperialism: Essays on Environmental History of South Asia*, New Delhi: Oxford University Press, 1995, p. 4.

[28] John R. McNeill and Alan Roe, eds, *Global Environmental History, An Introductory Reader*, London and New York: Routledge, 2012, p. xiii.

[29] *Ibid.*

[30] In explaining the concept of 'ecological nationalism', Cederlof and Sivaramakrishnan observe that at first glance ecology may be equated with

word ecology (*oecologie* in its first appearance, derived from a Greek term referring to the affairs of a family household) was coined by German biologist Ernst Haeckel in 1866, though the independent science of ecology, indispensable to environmental historians, did not exist for long after the turn of the century and ecological research attracted little attention except in the sciences. Most historians could not think of the environment except as Henry David Thoreau had, that is, less scientifically than philosophically or sentimentally. Notwithstanding the 'astigmatism of historians' then, shared by nearly all citizens of the industrialised nations, who could not accept their species as equal or subordinate to the other organisms in the biosphere,[31] there is no dilemma that man has been dependent on nature since time immemorial though the awareness of man 'as the maker and unmaker of nature' has been a more recent phenomenon and with it 'an awesome sense of our own capacity for mischief and mayhem'.[32] The ideology of environmental historians that appears largely biological emanates from the doubts of the choices that humanity has made, especially in more than a century, in exploiting the earth. The worry is about the durability of the intricate organic and inorganic relationships that support human beings.[33] The endeavour, consequently, has been towards telling the story of the intertwined fates of humanity and the natural world. Understanding the reciprocal relationship between nature and human

---

'ecologism' i.e. an attempt to preserve nature in complete terms without any discount to human beings. But that is not the issue, they argue. In fact, 'ecological nationalism' assumes a radical change in human relations with the environment, which may suggest changes in social and political life. In this sense 'ecologism' contrasts with environmentalism and this is why ecology may appear contrary to nationalism, the essential values of which are anthropocentric. Instead, they understand ecology, in its broad sense, as the inter-relatedness of the environment and organisms. See Gunnel Cederlof and K. Sivaramakrishnan, eds, *Ecological Nationalisms, Nature, Livelihoods and Identities in South Asia*, Ranikhet: Permanent Black, 2005, pp. 6–7.

[31] Alfred W. Crosby, 'The Past and Present of Environmental History', *The American Historical Review* 100: 4, October 1995, pp. 1177–89, particularly p. 1182.

[32] Arnold and Guha, 'Introduction', in *Nature, Culture, Imperialism,* p. 3.

[33] Crosby, 'The Past and Present of Environmental History', p. 1189.

beings, there has been a sharp increase in environmental concerns and activism leading to deep research interests in the twin stories of the human impact on the natural environment and the manifold influences of the environment on man.[34] Though a study of the environment in relation to man is sometimes viewed from the angle of 'nature versus nurture', it needs to be emphasised that each individual is the product of a certain genetic potentiality finding expression in a particular environment or a series of environments.[35]

The intellectual concern for the protection or conservation of nature goes back to the seventeenth and eighteenth centuries to the encounters of western European naturalists, doctors and administrators with the strange, unfamiliar tropical environments and the damage that they caused to these environments. From the mid-nineteenth to mid-twentieth century, environmentalism developed as historical geography, an idea reinforced by a symposium at Princeton, USA in 1955 that produced a two-volume anthology edited by William L. Thomas Jr., *Man's Role in Changing the Face of the Earth* (1956). This was also the phase when environmentalists, though without a substantial mass base, stressed upon the governments of North America and Europe to initiate the programmes for forest and water conservation and to establish national parks.[36] The environment movement in the USA was quite effective

[34] Mark Stoll, 'Series Forward', in Christopher V. Hill, *South Asia: An Environmental History*, Series: *Nature and Human Societies*, California: ABC-CLIO, Inc, 2008, p. xii; also pp. xvii-xviii of the same book; also see Paul E. Little, 'Environments and Environmentalisms in Anthropological Research: Facing A New Millennium', *Annual Review of Anthropology* 28: 1, 1999, p. 261; D. Worster, 'Doing Environmental History', in Worster, ed., *The Ends of the Earth*.

[35] Ranabir Chakravarti, 'The Creation and Expansion of Settlements and Management of Hydraulic Resources in Ancient India', in Grove et al, *Nature & The Orient*, p. 87.

[36] Ramachandra Guha, *Environmentalism, A Global History*, New Delhi: Oxford University Press, 2000, p. 3; Grove et al, *Nature & the Orient*, p. 3; also Richard Grove and Vinita Damodaran, 'Imperialism, Intellectual Networks, and Environmental Change, Origins and Evolution of Global Environmental History, 1676–2000', Part I, *Economic and Political Weekly* 41: 41, 14 October 2006, pp. 4345–54, especially pp. 4345–46; also John F. Richards, *The Unending Frontier*, pp. 19–20, 22; also see Richard Grove, *Green Imperialism: Colonial*

not as a result of government action, but due to the pressures of the environmentalists. It could thus protect its 'wilderness' from the risks or threats of development and control the disasters of industrialisation, that is, air and water pollution, the production of toxic or radioactive wastes.[37] American environmentalists could also successfully compel the American Congress to enact several environmental legislations. No longer a uniquely American privilege, the environment movement today represents a global interest in natural resource management, the desire among nations to collaborate, transcend boundaries and even compete but then the issue is: can it be characterised as a participatory and democratic or community-oriented system of natural resource management[38] or is the path to democratising nature 'littered with potholes' and does nature as an issue divide as much as it unites?[39]

Most American natural scientists trace the environment movement to the 1960s to Rachael Carson's books, namely *The Sea Around Us* (1952) and more specifically, *Silent Spring* (1962), that brought greater awareness on the threats of chemical pesticides and their impact on ecology and health. The latter was described as the Bible and the founding event of modern environmentalism. Carson transformed environmentalism from an elitist to a popular movement heralding 'New Environmentalism' and widening the audience for environmental history. A decade later, the book *Only One Earth* (1972) by the economist Barbara Ward and the geneticist Rene Dubois further cognised the issue.[40] Subsequently, Alfred Crosby setting a global agenda for environmental history

---

Expansion, Tropical Island Edens and the Origins of Environmentalism, 1600–1860, Cambridge: Cambridge University Press, 1995; also Crosby, 'The Past and Present of Environmental History', p. 1186.

[37] For a review, see Richard White, 'Historiographical Essay, American Environmental History: The Development of a New Historical Field', *Pacific Historical Review* 54: 4, 1985, pp. 297–335.

[38] Nicolas Laine and T. B. Subba, eds, *Nature, Environment and Society, Conservation, Governance and Transformation in India*, Hyderabad: Orient BlackSwan, 2012, p. xii.

[39] Rangarajan and Sivaramakrishnan, eds, *India's Environmental History, Colonialism, Modernity and the Nation*, vol. II, Ranikhet: Permanent Black, 2012, p. 15.

[40] The book by Ward and Dubois was released at the United Nations Conference on 'Human Environment' at Stockholm in 1972.

talked of 'ecological imperialism' to describe the successful ability of European colonialism to control resources worldwide. Classifying the industrial revolution as a capitalist scam and an ecological disaster that destroyed the native ecosystems and populations and paved the way for 'Neo-Europes', he spoke about the biological expansion of Europe in the New World while the 'Old World' civilisations like the Middle East, China and India though approachable remained resistant to the ecological imperialism of Europe. So, though the Europeans conquered the tropical countries that already had 'weary ecosystems', they could not Europeanise the tropics and thus had little impact on the ecosystems of these countries. To the contrary, by their 'portmanteau biota' (the term that Crosby uses for colonising the whites), the Europeans could Europeanise the temperate regions of North and South America and other Oceanic countries.[41] Pointing towards a glaring hiatus in Crosby's work, Richard Grove argued that the continents of Africa and Asia hardly figured in the Index of his book, let alone the main text and Crosby provided no answers as to what happened to environmental history outside North America.[42] However, even if one concedes to Crosby that the Europeans could not establish 'Neo-Europes' in India and China or that they could not decimate the population and natural resource base of these countries, it is pertinent nonetheless to observe that their intervention in the Indian society and ecology was so extensive and deep that it ultimately changed the ecology and the pattern of food production here. Exercising their political hegemonic powers, they attracted Indians towards their industrialism and consumerism ensuring that the process of ecological change that they had initiated would continue and intensify even after they had left India.

The vigour around environmental history in South Asia was noticed in the late 1980s.[43] The first attempt to review the history of resource

[41] Alfred Crosby, *Ecological Imperialism, The Biological Expansion of Europe, 900–1900*, New Edition, Cambridge: Cambridge University Press, 2004.

[42] Grove et al, *Nature & the Orient*, p. 5.

[43] Donald Worster equated environmentalism with the conservation movement asserting that conservation of forest lands constituted an early phase of environmentalism—even when those lands were set aside for timber supply and revenue. Only after World War II did the focus of the environmental movements shift to pollution and health concerns. See his

management, particularly the forests[44] in the region was made by B. Ribbentrop in his book *Forestry in British India* (1899) followed by E. P. Stebbing's three volumes, *The Forests of India* (1922).[45] In environmental social sciences, forests have held precedence over other ecosystems, partly being a vast natural resource and placed early within the ambit of colonial 'utilitarian scientific experiments'.[46] The two British forest officials argued strongly in favour of the benefits of scientific forestry, scientific supervision and management of forests. They contended the necessity of 'reserved' forests to protect them from hazardous private interests though access to the limited tracts of forests was allowed to the villages. The foresters of the British Empire usually claimed to understand the broad implications of their work and the effect of the practice of forestry on the environment. In 1872, Baden Powell admonished his officers at the first India-wide forestry conference to 'regard the planting and restoration of our divisions as your chief business'. He instructed: 'never consent to work as if the felling of timber was the great work of life and as if the provision of a few rupees in the Budget under the planting head…was all that is needed by way of supplement'.[47] Observations such as these by British surveyors persuade one to ask whether the so-called newly protected forests marked the beginning of the formalisation of a divorce that had been threatened since the early modern period, that is, the final transition from man-in-nature to man-over-nature.[48]

American Environmentalism: The Formative Period, 1860–1915, London: John Wiley & Sons, 1974, pp. 85–95.

[44] Gregory A. Barton argues that the fear of human engendered catastrophe focused early on forests. Revenue from timber but also soil erosion, flooding and rainfall perhaps fed these apprehensions. See his *Empire, Forestry and the Origins of Environmentalism*, Cambridge Studies in Historical Geography 34, Cambridge: Cambridge University Press, 2002, p. 28.

[45] For more details on Stebbing and his work, see Grove and Damodaran, 'Imperialism, Intellectual Networks, and Environmental Change', Part I, p. 4349.

[46] Cederlof and Rangarajan, 'Predicament of Power and Nature in India: An Introduction', *Conservation and Society* 7: 4, 2009, pp. 221–26, in particular, p. 223; also Barton, *Empire, Forestry and the Origins of Environmentalism*, p. 10.

[47] Baden Powell and J. C. Macdonell, *Report of the Proceedings of a Conference of Forest Officers* held at Lahore, 2–3 January 1872, p. 90, cf. Barton, *Empire, Forestry and the Origins of Environmentalism*, p. 9.

[48] Barton, *Empire, Forestry and the Origins of Environmentalism*, p. 37.

Contrary to the views of Ribbentrop, Stebbing and other British forest officials, many contemporary Indian scholars like Vandana Shiva, Ramachandra Guha and Madhav Gadgil have argued that the colonial forest policy led to deforestation and that 'every day beliefs of tribal and peasant society' were smashed by the commitments of the colonial government.[49] The colonial government, urges Ramachandra Guha, manipulated the customary use of forests to enable sustained timber production fulfilling their commercial and strategic interests, leading to ecological decline and agrarian protest. The British, Guha reiterates, were the world's premier 'omnivores', drawing resources of the entire biosphere to their tiny island kingdom and meeting the resource consumption needs of the British elite by draining their colonies including India of their natural resources. Stipulating on the 'organic' connection between industrialisation and imperial expansion as the European colonists occupied large parts of the world and reoriented local economies to cater to the needs of the metropolis, Guha observes that decimating the forests through the eighteenth and the nineteenth centuries, not only in India but also north-eastern USA and southern Africa, the British were 'unquestionably the world leaders in deforestation'. He observes that all European powers—the Dutch, the Portuguese, the French, the Belgians and the Germans—emulated the British and became the 'prime agents of ecological destruction in their colonies'.[50] Applying the concept of the environmentalism of the poor developed by Martinez-Alier, he mentions situations that have pitted rich against poor, logging companies against hill villagers, dam builders against forest tribals, and multi-national corporations deploying trawlers against artisanal fisher folk rowing country boats.[51] Elizabeth

[49] Vandana Shiva, *Staying Alive: Women, Ecology and Survival in India*, New Delhi: Kali for Women, 1988, pp. 54–59.

[50] Ramachandra Guha, *The Unquiet Woods: Ecological Change and Peasant Resistance in the Western Himalaya*, New Delhi: Oxford University Press, 1989, pp. 8, 55; also his *Environmentalism*, p. 4; also Madhav Gadgil and Ramachandra Guha, *Ecology and Equity, The Use and Abuse of Nature in Contemporary India*, New Delhi: Penguin Books, 1995, p. 9.

[51] Ramachandra Guha, 'The Environmentalism of the Poor', in R. G. Fox and O. Starn, eds, *Between Resistance and Revolution*, New Brunswick: Rutgers University Press, 1997, pp. 19–20; also Ramachandra Guha and J. Martinez-Alier, *Varieties of Environmentalism, Essays North and South*, London:

Whitcombe, in her work on the agrarian conditions of the United Provinces during the British rule in the nineteenth, century heralded these arguments. She observed that the vast irrigation investments made by the English East India Company in the Indus and Ganges basins subsequent to 1820 CE proved ecologically disastrous and caused widespread salinisation and *reh* (a salty surface crust found on the soil) deposition rather than economic development.[52] Reinforcing these arguments, but in a different context, Laxman Satya, in his pioneering ecological history of cattle, observes that it was the environmental degradation procreated by the colonial state that led to the ecological crisis and spelled doom for the cattle-rich zone of Berar in central India. The region experienced an abnormal ecological flux that disrupted the regular annual *kharif* (autumn harvest) and *rabi* (spring harvest) cropping pattern. He rejects the colonial attempt to explain famine and disease as a natural outcome of the backward tropical environment and dismisses it as 'not only too simplistic and prejudicially naive but a part of the imperialist ideology of conquest and domination'.[53]

Emphasising on the conflict between state forestry and peasantry and environmental degradation, the above-mentioned academe have enforced a pattern of uniformity in the sphere of forestry though there is sufficient space to ponder on elements of diversity between regions. They have suggested that the global capitalist economy and the colonial commercial policy in South Asia were the main culprits for the depletion of the world's natural resources that led to deforestation, destroyed the ageless harmony between people and nature, and alienated the communities residing in the forest areas by depriving them of the right to manage the forests. Such literature, no doubt, favoured grassroot activism and empowerment of the local community but it ignored or underrated the fact that the peasant or tribal groups had a long history of successive adaptations to change; they were not 'forlorn

Earthscan Publications, 1997; also J. Martinez-Alier, 'Ecology and the Poor: A Neglected Dimension of Latin American History', *Journal of Latin American Studies* 23: 3, 1991, pp. 621–39.

[52] Elizabeth Whitcombe, *Agrarian Conditions in Northern India: The United Provinces under British Rule, 1800–1869*, Berkeley: University of California Press, 1972.

[53] Laxman D. Satya, *Ecology, Colonialism and Cattle: Central India in the Nineteenth Century*, New Delhi: Oxford University Press, 2004, p. 182.

relics of an ecologically friendly golden age', neither were they socially homogenous but strongly differentiated and stratified according to economic standing, local power, gender and/or any other factor.[54]

Historians like John Richards, Edward Haynes, Elizabeth Flint and Jim Hagen have analysed the changes in land use over time and have dispelled the myth that there was serious desertification in India's arid and semi-arid zones. Richards assessed the nuances of environmental history and observed that if human-induced environmental change is seen as 'an unrelieved tragedy of remorseless ecological degradation and accelerating damage', it underestimates the resilience of ecosystems and overestimates human-induced impacts as opposed to natural processes for example, in soil erosion.[55] Similarly, scholars like Richard Grove have challenged the idea of desiccation that has dominated tropical environmentalism since the nineteenth century and have demolished the notion that human action destroyed forests. Grove urges that the investigation into the motives underlying the emergence of environmentalism and conservation policies in the colonial context points to a 'heterodox legacy' of Utopian, Physiocratic and Hippocratic philosophical elements represented in the discourses of an influential minority of scientists, who were distinctly connected to a variety of agendas of social reform, guided by the incentives of agrarian prosperity and social stability.[56] Agreeing with Grove that the hypothesis of 'environmental imperialism'[57] does not stand up well at all, Marlene Buchy, through his research on colonial forest policy in the district of

[54] Laine and Subba, eds, *Nature, Environment and Society*, p. xiv; also McNeill and Roe, eds, *Global Environmental History*, p. xiv.

[55] Richards, *The Unending Frontier*, p. 13.

[56] Grove, *Ecology, Climate and Empire*, pp. 84–85; on desiccation, see Grove, *Green Imperialism*, especially pp. 346–77. Similar arguments have been made by James Fairhead and Melissa Leach for West Africa. Their research challenges the idea of desiccation that has dominated tropical environmentalism since the nineteenth century and has demolished the notion that West African forests are but the miserable remnants of vast woodlands destroyed by human action. See James Fairhead and Melissa Leach, *Misreading the African Landscape: Society and Ecology in a Forest-Savannah Mosaic*, Cambridge: Cambridge University Press, 1996.

[57] Richard Grove, 'The Origins of Environmentalism', *Nature* 345, 3 May 1990, p. 14.

North Canara (now part of Karnataka), shows that colonial forestry was not so much ecologically destructive as 'unadapted' and therefore socially and (to some extent) economically disruptive. He further stipulates that modern concepts such as sustainability or biodiversity are difficult to apply in a *post hoc* sense.[58] Sumit Guha, however, has focused on ethnicity and environment and emphasises on the need to study the role of human activity in the forest zones to consider their impact.[59] Ghazala Shahabuddin, on the other hand, is critical of all such political ecologist and environmental historians who stress on the political exclusion of people from natural resources and the colonial subversion of the conservation agenda or the belief that the creation of people-free wilderness such as national parks is the best way to protect nature.[60] Such impressions, she argues, tend to overlook the more local causes of ecological degradation. Reiterating her stance, she insists that there are no easy answers either to the question of the causes of ecosystem erosion or of the best ways ahead.[61]

From 1982 onwards, environmental history, particularly forest history in India, became intensely politicised. The process of politicisation began with the Chipko movement in the Garhwal region, Uttarakhand—a peasant initiative against commercial forestry that combined the issues of the poor and the marginalised with gender. Chipko demonstrated the resilience and folk wisdom of the rural poor, represented real history of protest against state forestry in India and was connected to the broad history of national protest and freedom struggle.[62] Providing an ecological dimension to the study of agrarian

[58] Marlene Buchy, 'British Colonial Forest Policy in South India: An Unscientific or Unadapted Policy?', in Grove et al, *Nature & the Orient*, pp. 637, 669.

[59] See Sumit Guha, *Environment and Ethnicity in India, 1200–1991*, Cambridge Studies in Indian History and Society 4, Cambridge: Cambridge University Press, 1999.

[60] See Ramachandra Guha, *Enviornmentalism*; also A. Chhatre and V. K. Saberwal, *Democratizing Nature: Politics, Conservation and Development in India*, New Delhi: Oxford University Press, 2006; also Gadgil and Guha, *This Fissured Land*.

[61] Shahabuddin, *Conservation at the Crossroads*, pp. xviii–ix.

[62] Apart from Chipko, the 1970s witnessed a number of popular movements in defense of local rights to forest, fish and water resources as well as protests

history and peasant resistance, Ramachandra Guha argues that ecological history cannot merely be the history of changes in the landscape; it must link environmental changes with changing and competing human perceptions of the 'uses' of nature.[63] For Sunita Narain, however, with Chipko emerged the environmental challenge to issues of access and sustainable management of natural resources. Women had prevented timber merchants from cutting the forest not merely to protect the forest but to assert their right to the resources of the forest; they knew that they had to protect the forest to protect their livelihood. It was a call to redefine development and growth.[64]

Women's environmental movements emerge when gender becomes a determining factor in issues involving the division of labour, access to natural resources and property relations in ways that are disadvantageous to women. Women's resistance movements directly related to environmental issues has generated new fields of feminist political ecology and eco-feminism. Defining 'Feminist Environmentalism', Bina Agarwal observes that women's and men's relationship with nature should be understood as 'rooted in their material reality' and in their specific forms of interaction with the environment. For example, the link between women and the environment can be seen as structured by a given gender and class, caste or race organisation of production, reproduction and distribution. Agarwal argues that women, especially in poor rural households in India, have been victims of environmental degradation in gender-specific ways. But, on the other hand, giving themselves a gender-specific perspective, women have also been active agents in the environmental movements for protection and regeneration. For instance, poor peasant and tribal women have characteristically been responsible for fetching fuel and fodder; in hill and tribal communities,

---

against large dams. These movements occurred in India, Brazil, Malaysia, Ecuador and Kenya among peasants, pastoralists and fishermen. For more details and a comparative analysis of these movements, see J. Martinez-Alier, *The Environmentalism of the Poor*, London: Edgar Elgar, 2004.

[63] Guha, 'Introduction', *Social Ecology*, p. 1; also his *The Unquiet Woods*, pp. xiii–iv; also Gadgil and Guha, *Ecology and Equity*, pp. 84–85.

[64] Sunita Narain, 'Editorial – India's Twin Environmental Challenges', in *Fortnightly News Bulletin,* Centre for Science and Environment, New Delhi, 3 December 2013.

they have also often been the main cultivators. They are thus most prone to be adversely affected in quite specific ways by environmental degradation. At the same time, in the course of their daily interactions with nature, they acquire a special understanding of species, varieties and the process of natural regeneration. (This would include the lessons passed on to them, for example, by their mothers.) They could thus be seen as both victims of the destruction of nature and as repositories of information about nature, in ways that are distinct from the men of their class. In case of environmental devastation, therefore, the gendered impulse and response would be one of resistance but their knowledge of the environment, on the other hand, would condition their perceptions and choices of what should be done. On the basis of their experiential perceptions, they could provide an insight and a direction on the processes of environmental replenishment, which may offer alternative approaches to development. This implies that if women abandon their contact with the natural environment or if they do not actively implement their environmental wisdom in their daily sustenance, they may lose their knowledge over time and the possibility of its transmission to others.[65] The gender-specific roles in the environmental context are also evident in Vandana Shiva's understanding of colonialism and the evolution of masculinist forestry. She argues that the British not only displaced local rights, but also ignored the local needs and local vision of the people to sustainably manage the forests and 'reduced this primary source of life into a timber mine'. The commercial use of forests and the ruination of the forest ecosystem deprived women of their subsistence economy; in fact, it adversely affected all such groups, women and tribal communities included, who depended on the diverse resources of the forests for their livelihood.[66]

It is obvious that ecological and socio-economic linkages between the impacts of forest laws, mining, dams and occupational hazards

[65] Bina Agarwal, 'The Gender and Environment Debate: Lessons from India', in Nivedita Menon, ed., *Gender and Politics in India*, Themes in Politics, New Delhi: Oxford India Paperbacks, 2004, pp. 96, 105–06; also see Bina Agarwal, *A Field of One's Own: Gender and Land Rights in South Asia*, Cambridge: Cambridge University Press, 1994.

[66] Vandana Shiva, 'Colonialism and the Evolution of Masculinist Forestry', in Nivedita Menon, ed., *Gender and Politics in India*, pp. 39–41.

have led to growing anxiety regarding issues of the environment and ecological disasters alongside 'Green' movements.[67] More significantly, *adivasi* (tribal) and peasant movements concerning the community rights to forests has aroused historical curiosity about colonial forest policies and led to popular protests, subsequently extending to the politics of water use and big dams. Despite the existence of enough evidence of environmental degradation, the centralisation of authority has failed to protect the environment. It is pressing that environmental politics be understood in terms of 'who benefits' and 'whose interests are being protected'.[68]

The perceptions and apprehensions of the environmental crisis have engulfed the world with growing industrialisation that has threatened to transform the *natural order* by new methods of resource extraction, production and transportation. The scale and intensity, observes Ramachandra Guha, at which nature is used or abused has increased in several ways. Medical technology has resulted in a steady improvement of lifespan and increase in human population. As more humans produce more and consume more, it accelerates environmental hazards, resulting in greater pollution and loss of habitat. The environment has become a source of cheap raw material and a dump-yard for the unwanted residues of economic growth; constantly growing urge for mining and industry have decimated the forests and the habitats of wild life; dangerous and risk-prone chemicals have been excreted into the rivers and the atmosphere.[69]

However, with significant changes throughout the world in the inter-relationship between man and his environment caused by rapidly expanding urbanisation and industrial growth that can irreversibly destroy natural ecosystems, there is also a constant engagement with the issues of how to strike a balance with nature or establish compatibility between growth and equilibrium; preserve spaces for nature even as development continues or whether and where the fertile, potential lands

[67] Sumit Sarkar, *Modern Times, India 1880s–1950s*, Ranikhet: Permanent Black, 2014, p. 77.

[68] See Dunu Roy, 'Environmentalism and Political Economy', *Seminar 516*, August 2002, pp. 65–68, particularly p. 66.

[69] Ramachandra Guha, *Environmentalism,* pp. 3–5; also see his *How Much Should a Person Consume? Thinking through the Environment*, Ranikhet: Permanent Black, 2006, pp. 5–6; also, Gadgil and Guha, *This Fissured Land.*

can be replaced by industrial sites.[70] Growth and equity, in fact, have become two major political issues in recent times and the environment movement has added a third one, namely, sustainability. The issue of sustainability has raised serious questions about the kind of growth we want, re-emphasising the need for equity, both globally and locally.[71] Despite a global consensus on these matters, the western industrialised nations including the erstwhile colonial powers have remained hesitant in their commitment to significant reductions or compensation for 'historical offences' or pledging costs for necessary improvements pitting issues of environment against historical legacies, historical justice and a desire for development.[72]

III

The focus on the notion of *agrarian environment* underscores the fact that while environmental history has made significant contributions to the understanding of the inter-braided, intertwined human–nature relationship, its self-imposed constraints and its isolation from agrarian history are both problematic and conceptually unacceptable. Environmental history predicated on a critique of scholarship and politics that was engaged in agrarian and urban studies, considered that progress through agrarian expansion led to deforestation and thus expelled the farms and agricultural fields from its purview in the same way that agrarian history had earlier neglected the forests and pastures. It was 'as if these histories were not connected, as if environment did not impinge on the agrarian world'.[73] In this context, it is requisite

[70] Cederlof and Rangarajan, 'Predicament of Power and Nature in India', p. 222; also Shahabuddin, *Conservation at the Crossroads,* p. xiv.

[71] Anil Agarwal, 'An Indian Environmentalist's Credo', in Guha, ed., *Social Ecology*, p. 347.

[72] Cederlof and Rangarajan, 'Predicament of Power and Nature in India', p. 222.

[73] Neeladri Bhattacharya, 'Introduction', Special Issue 'Forests, Fields and Pastures', *Studies in History* 14: 2, July–December 1998, pp. 165–71; also Arun Agrawal and K. Sivaramakrishnan, eds, *Social Nature, Resources, Representations, and Rule In India*, New Delhi: Oxford University Press, 2001, p. 2; also Sarkar, *Modern Times*, pp. 78–79.

to assert that instead of upholding the social identities related to the human–nature relationship that confirm the separation of the agrarian and environmental as independent domains and choosing between an agrarian or an environmental perspective is a misrepresentation of the interwoven dynamics between the agrarian and the environmental worlds.[74] It is vital to remove the rift between the environmental and the agrarian, argue for a more nuanced and dialectical relationship between the two and urge upon 'an *ecologicalism* that recognizes the inherent interdependence of all life-systems'.[75] Agrarian environments have to be understood as being part of a biophysical and social environment; urban and the non-urban; arable and non-arable; and areas that are integrally connected to the realm of agriculture and environment and their allied socio-economic relations.[76]

The broad definition of agrarian environment is an essential prerequisite because a variety of specialised kinds of production are knitted in a tight web in agrarian environments urging the need for a term that can encapsulate different specialists even as their situations are considered separately. The importance of agrarian environments can be further demonstrated by focusing on landscapes of social power and bringing together the various dimensions of agriculture. Agrarian history is not just a local matter. It is replete with examples of ecological interventions whereby nature has been altered and landscape modified to fulfil the expectations of the people. Agriculture was and continues to be the single most important means by which mankind has changed the lands and the ecosystems. It represents a connecting point between the human powers that organise agriculture and the changing natural environment and is a major element of ecological transformation in human history, for no occupation other than farming alters the land so much. New cultivation occurred in the frontiers of settlement, when the pioneer settlers intruded upon the remote lands inhabited by shifting cultivators, hunter-gatherers and pastoralists. Agrarianism, nevertheless, includes not only farming but also animal husbandry, pastoralism, fishing

[74] Agrawal and Sivaramakrishnan, eds, *Social Nature*, pp. viii, 1–2, 5.

[75] Daniel Esty and Marian Chertow, eds, *Thinking Ecologically: The Next Generation of Environmental Policy*, New Haven: Yale University Press, 1997, p. 45.

[76] Agrawal and Sivaramakrishnan, eds, *Social Nature*, p. 6.

and harvesting the forest. It can thus be defined as a major source that produces organic material for human use, of which animal and forest products are evidence.[77]

For a distinct, discrete understanding of agrarian environments, it is worthwhile to investigate the varied opinions on the effects of human intervention with nature. Conventionally, land is considered to be upgraded if output increases. But when it brings reduced biodiversity and biomass, ecological perspective sees it as degradation, and a decline of land value. 'Technological optimists' have, however, challenged the 'eco-doomsters' and denied the existence of any problem of degradation. They have argued that land has been made more useful by man's efforts, knowledge and technological advances. Consequently, world societies have not only been able to meet the rising demand of their increasing population but have also moved towards growing commercialisation and industrialisation. The 'eco-alarmists', on the other hand, have failed to view degradation within a wide historical and geographical framework. They have projected all changes in the environment as irretrievable damage. The analysis of these two dichotomous views conveys that constant change occur in landscapes and many cause degradation, but then, it is equally important to concede that the term degradation is a matter of perception and is likely to be diversely used.[78] For instance, sedentary cultivation and grazing are considered negative activities that invariably lead to the erosion of the soil. Grazing is deemed dangerous to a forest because as the sheep and the cattle eat the seedlings of trees, they damage its reproductive potential. Yet, the use of graze was approved by colonial foresters like Dietrich Brandis, Ribbentrop and

[77] See John F. Richards, 'Only a World Perspective is Significant: Settlement Frontiers and Property Rights in Early Modern World History', in Jill Conway, Kenneth Keniston and Leo Marx, eds, *The Humanities and the Environment*, Cambridge: MIT Press, 1998; also Richards, *The Unending Frontier*, pp. 4–10; also Ludden, *Agrarian History*, pp. 16, 18, 49, 60.

[78] Meena Bhargava, *State, Society and Ecology: Gorakhpur in Transition, 1750–1830*, Revised Edition, New Delhi: Primus Books, 2014, p. 33; For more details, see John F. Richards, 'World Environmental History and Economic Development', in W. C. Clarke and R. E. Munn, eds, *Sustainable Development of the Biosphere*, Cambridge: Cambridge University Press, 1986, pp. 53–74; also see Piers Blaikie and Harold Brookfield, eds, *Land Degradation and Society*, London and New York: Methuen & Co., 1987.

Stebbing probably since grass formed and continues to form the basis of Indian peasant economy. That may also have been the reason why the colonial forest department felt compelled to confirm the grazing rights of the peasants before attempting its plan of 'reserved' forests.[79] The pastoralists, thus, may not reckon the replacement of forests by grasslands as degradation even though the hoofs of large number of cattle driven into a forest to graze may make the ground hard, compacted and dry, rendering it less conducive for the seeds to germinate or plants to grow.[80] Similarly, the farmers may not perceive the replacement of forests by agricultural lands as degradation despite the adverse changes in the soil and micro-climate;[81] and logging that may restrain the new plant growth.[82] Moreover, the farmers often prefer agriculture and the development of cultivation in the proximity of or within the ambit of a forested area so as to benefit from the springs or a high water-table or less eroded soil or even the new soil produced by the decayed leaf litter in these areas.[83] In several regions, the debris of burnt forest waste was considered a particularly good medium for seed growth. An abundant supply of such waste was well-suited to agriculture—a feature commonly noticed in the forest tracts and wastelands (uncultivated and/or barren lands) of Mughal *sarkar* (sub-division of a province)Gorakhpur in *suba* (province) Awadh.[84]

The congeniality and harmony between the agrarian and the environmental are evident by the fact that while agriculture developed leading to a sedentary lifestyle, forests remained a way of life for many of its inhabitants. It served as their home, a shelter, a source of fodder, firewood and timber even if the capital value of the latter was of no great importance to them. The *zamindars* (landlords) possessed proprietary, hereditary rights over the forests within their jurisdiction. Generally,

[79] Richards, *The Unending Frontier*, p. 11; also Barton, *Empire, Forestry and the Origins of Environmentalism*, pp. 89–90.

[80] Shahabuddin, *Conservation at the Crossroads*, p. x; also Laeeq Futehally, *Our Environment*, repr., New Delhi: National Book Trust, 2004, p. 33.

[81] Micro-climate refers to a set of climatic conditions, measured at a very local level, which directly affects the growth and reproduction of animals and plants. See Shahabuddin, *Conservation at the Crossroads*, p. x.

[82] *Ibid.*

[83] Futehally, *Our Environment*, p. 33.

[84] Bhargava, *State, Society and Ecology*, p. 34.

they did not consent to the cutting of trees nor did they, as Buchanan Hamilton observed, 'agree to receive a pecuniary compensation for this sacrilege'.[85] Large parts of forest were home to large mammals, ungulates and carnivores especially lions and tigers, cheetahs and leopards, hyenas and wolves, elephants and ants. Some ant hills, for example in Mughal sarkar Gorakhpur, were found to be at least 10 feet high; many trees were coated with ant tunnels and a few points of ant nests were broken with large open galleries inside. However, irrespective of whether the forests were dense or subjected to the incursions of wild animals that made the passage through the forests dangerous and difficult, there was nothing to deter the local inhabitants from defining their relationship with the forests.[86] Significantly, there was nothing rigid in the forest line. Lands were reclaimed; agro-forestry, pastoralism and settlements could overwhelm the forests. There was constant fluctuation and fluidity between forests and cultivation, with shifting boundaries between the jungle and the arable. Fath Ullah Shirazi's observation during the reign of Akbar regarding revenue matters may be quoted as substantiation: 'The fluctuations of civilization are apparent to everyone. If in a village some land falls out of cultivation one endeavours to increase cultivation elsewhere'.[87] The pre-colonial rulers had encouraged reclamation of forests by granting *nankari*[88] rights to the landholders. The Sultans of Delhi, for instance,

[85] Buchanan Hamilton, 'An Account of the Northern Part of the District of Gorakhpur', Book III, Mss Eur D 91–93, India Office Library, London, p. 55. Also see David Hardiman, 'Power in the Forests: The Dangs, 1820–1940', in David Arnold and David Hardiman, eds, *Subaltern Studies VIII*, New Delhi: Oxford University Press, 1994, p. 105.

[86] Bhargava, *State, Society and Ecology*, pp. 36–37; also her 'Forests, People and State: Continuities and Changes', *Economic and Political Weekly* 37: 43, 26 October 2002, pp. 4440–46. This article was also published in Meena Bhargava, ed., *Exploring Medieval India, Sixteenth to Eighteenth Centuries, Politics, Economy, Religion*, vol. I, Hyderabad: Orient BlackSwan, 2010, pp. 287–306.

[87] See Chetan Singh, 'Forests, Pastoralists and Agrarian Society in Mughal India', in Arnold and Guha, eds, *Nature, Culture, Imperialism*, pp. 34–35. (also chapter 3 of this volume).

[88] *Nankari* or *nankar* was an allowance for *khidmat* or service performed by the zamindars in collecting and remitting land revenue. It was granted in the form of a deduction from the revenue paid or in the form of revenue-free land.

had transformed the spaces—deserted lands and jungle of thorn and scrub—around the Qutb complex in Delhi into settled areas through the thirteenth century. Sultan Balban's campaigns in the vicinity of Delhi involved clearing the forests. He mustered a large number of *tabarzan* (woodcutters) from amongst his soldiers to clear the paths of the army. Balban also ordered the resettlement of the reclaimed lands and rewarded those who facilitated this by granting them revenue-free lands.[89] Similar connections between forests and military campaigns and creation of settlements were noticed during the Mughal period. From the reign of Emperor Shah Jahan, the Mughal army was accompanied by woodcutters and ploughmen to clear the land to make way for the army and establish new *zamindaris*. Asserting their political power, the zamindars and the *jagirdars* (land revenue assignees) too, hired professional woodcutters to clear the jungles and settle new villages. In the late eighteenth century, for example, the zamindars of Mughal sarkar Gorakhpur, who were the sole holders of timber rights, planted *sal* (known as *sakhuya* in Gorakhpur) and allowed it to be cut and sold. With such incentives and entrepreneurship, many zamindaris emerged at the frontiers of cultivation consequent upon the clearings of forests.[90] While reclamation was carried out in some parts of this Mughal sarkar in 1801, there were tracts of jungles and wastelands that remained along river Gandak and its numerous streamlets that were used as *dhab* lands (pasture lands) during cold and hot months.[91]

Forests, pastures and cultivated fields served as resources to be exploited by specialists and it were the specialisations that constituted

[89] Sunil Kumar, *The Emergence of the Delhi Sultanate*, Ranikhet: Permanent Black, 2007, pp. 324–25, 333–36.

[90] Muzaffar Alam, 'Eastern India in the Early Eighteenth Century "Crisis": Some Evidence from Bihar', *Indian Economic and Social History Review* 28: 1, 1991, p. 66; also his book *The Crisis of Empire in Mughal North India: Awadh and the Punjab, 1707–48*, New Delhi: Oxford University Press, 1986, p. 98; also see Satish Chandra, 'Role of the Local Community, the Zamindars and the State in Providing Capital Inputs for the Improvement and Expansion of Cultivation', in Satish Chandra, *Medieval India: Society, Jagirdari Crisis and the Village*, New Delhi: Macmillan, 1982, p. 175; also Chetan Singh, 'Forests, Pastoralists and Agrarian Society in Mughal India', chapter 3 of this volume; also Bhargava, *State, Society and Ecology*, p. 36.

[91] Bhargava, *State, Society and Ecology*, pp. 35–36.

the various communities. Yet, specialised roles could not support entire communities and most people depended on multiple activities to sustain themselves as is evident from the example of the *bhil*s of Khandesh around 1850 CE.[92] So, while it would be prudent to consider the impact of human activity on the fauna and the flora that have fairly changed the ecosystem, it would be equally pertinent to observe that nature's ability to reconcile competing human interests is not limitless nor are all humans equally culpable of exploiting nature.[93] In this context may be mentioned the nomadic pastoralists who have been cited as predators to the sedentary agricultural societies, yet the two major forms of subsistence—pastoralism and agriculture—are neither totally separated nor fixed, but interchangeable and interactive,[94] with dynamic interaction between the two.[95] There are occasions when specialised herding is often made possible only by a symbiotic relationship with agriculture. Agro-pastoralism prevails in zones that cannot exclusively support successful agriculture or pastoralism; those who earn their livelihood largely from herds spend less time and labour on fields and vice versa.[96] In a situation

[92] The bhils derived their livelihood and sustenance by either cultivating or cutting and selling grass, firewood and timber, gathering and selling jungle fruits, roots, herbs, etc or during harvest season they watched the grain-fields of the farmers or reaped their crops but there were still others who took to the plough permanently whereas many took to hunting and killing wild animals with the hope of an award from the British government. See Sumit Guha, *Environment and Ethnicity*, pp. 28, 41.

[93] Cederlof and Rangarajan, 'Predicament of Power and Nature in India', p. 223.

[94] Paul E. Little, 'Environments and Environmentalisms in Anthropological Research', pp. 262–63; also Xinru Liu, 'Migration and Settlement of the Yuezhi-Kushan, Interaction and Interdependence of Nomadic and Sedentary Society', in Xinru Liu, ed., *India and Central Asia*, Ranikhet: Permanent Black, 2012, pp. 73, 102; also see Agrawal and Sivaramakrishnan, eds, *Social Nature*, pp. 5–6.

[95] Ibn Khaldun, Arabic historian of the fourteenth century, emphasises on the importance of dynamic interaction between the sedentary and the nomadic peoples to analyse the process of state formation. See Ibn Khaldun, *The Muqaddimah: An Introduction to History*, trans. F. Rosenthal and N. J. Dawood, London: Routledge & Kegan Paul, 1987; cf. Gommans, *Mughal Warfare*, p. 3.

[96] Shereen Ratnagar, *The Other Indians, Essays on Pastoralists and Prehistoric Tribal People*, New Delhi: Three Essays Collective, 2004, pp. xi, 97–98.

where pastoral-nomadism thrives, a movement towards sedentarisation is possible, like it was in the case of medieval Punjab, where the growth of commercialised agriculture under the Mughals encouraged and facilitated the absorption of pastoral people into the agrarian system. The contribution of the sedentary communities, on the other hand, to the creation and consolidation of economy and society cannot be seen as isolated or exclusive of the tribal communities. In fact, there are several instances of interdependence between tribal and sedentary life; one such example being the village communities of Mughal India that subsisted at a reasonable proximity to tribal societies. There was continued participation of tribal populations in the socio-economic processes and the stabilisation of agrarian society and certain other social formations, although in many cases, the sedentarised relied considerably on pastoralism. Even within the cultivated area there existed not only variations of the agrarian system but also socio-economic systems that were remarkably different from it. There are innumerable examples from several regions that demonstrate the close relationship between tribal-pastoralism and sedentary-commercialised society and so inextricably linked are its economic interests that any shift in the structure of any one of these societies would not leave the other unaffected. Chetan Singh argues that if tribal-pastoral populations in other regions were as integral to the agrarian society as in medieval Punjab then ignoring them would mean risking an incomplete understanding of the agrarian system of Mughal India—an understanding that need not be confined to the Mughal period alone.[97] In some instances, however, the reverse was evident with cultivators taking to pastoralism; in yet other circumstances, *jhum* (shifting cultivation) cultivators moved to settled cultivation and the cultivators escaped to the forests. On several occasions forests served as 'hideouts' and provided a haven for such peasants and zamindars, who escaped to the jungles to resist and avoid payment of taxes to the Mughal state. Sujan Rai described the Lakhi jungle near the channels of rivers Sutlej and Beas as an area of refuge for the recalcitrant in the late sixteenth century (1595–96 CE) while the *Tuzuk-i Jahangiri* informs

[97] Chetan Singh, 'Conformity and Conflict: Tribes and the "Agrarian System" of Mughal India', *Indian Economic and Social History Review* 23: 3, 1988, pp. 319–40. This article also reproduced in Bhargava, ed., *Exploring Medieval India*, vol. I, pp. 259–86.

us that during the reign of Jahangir, villagers in the vicinity of Mathura fled to the jungles on the other bank of river Yamuna to resist payment of land revenue. Forests served as an asylum too for the seriously ill and convalescents of infectious diseases, particularly smallpox, rampant in the eighteenth century in several areas. The disease was so dreaded that infected persons were removed from their homes and sent to the forest until the disease was eradicated from the village.[98]

Community locations, thus, were rarely fixed during the medieval–early modern period when desertions and migrations were often the response to climatic or political factors resulting in situations where wanderers might settle, and the settled wander; forests be cleared and forests be grown.[99] In other words, the shift from one form of livelihood to another was neither unilinear nor complete, nor did they exist in an isolated, exclusive, pure form.[100]

IV

To put the medieval–early modern environment in a perspective in the context of these debates, essays for the volume have been selected that investigate the different constituents of the environment, intrinsic to each other, and assess human–environment interaction. In the section on 'Nature and its Perception', the two essays by Salim Ali, and M. A. Alvi and A. Rahman discuss the Mughal emperors' varied but controversial, contradictory perception of nature. While the Mughal emperors exhibited tremendous passion for nature, were keen on researching the fauna and flora and maintained a large royal menagerie, they were also incorrigible, obdurate hunters with no apparent demonstrated concern

[98] Alam, 'Eastern India in the Early Eighteenth Century "Crisis"', p. 66; also his *The Crisis of Empire*, p. 98; also Shireen Moosvi, 'Man and Nature in Mughal Era', Symposia Paper 5, Thematic Symposium 'Man and Environment in Indian History', *Proceedings of the Indian History Congress*, 54th Session, Mysore, 1993, p. 13; also see Meena Bhargava, 'Forests, People and State', in Bhargava, ed., *Exploring Medieval India*, vol. I, pp. 294–95.

[99] Sumit Guha, *Environment and Ethnicity*, pp. 28–29, 41–42, 55.

[100] Andre Wink, *Al Hind: The Making of the Indo-Islamic World, Early Medieval India and the Expansion of Islam, 7th–11th Centuries*, vol. 3, New York: E. J. Brill, 1996, pp. 81–82, 93.

for the preservation of species. Salim Ali[101] mentions the frequent hunting expeditions of Jahangir, when from age 12 to 50, he had hunted 28,532 animals of which 17,167 were killed by the emperor himself; Shah Jahan's hawking and hunting with the cheetah and Aurangzeb's fondness for chase and hunting the lion. With figures such as these, one is constrained to speculate the role of the Mughal emperors in the decline of these species, especially in view of the hunt of the rhinoceros[102] by Babur and Humayun. As opposed to Salim Ali, Alvi and Rahman in their 'Introduction' (included in the volume as chapter 2, titled '*Jahangir – The Naturalist*') to *Jahangir – The Naturalist* present Jahangir as a positive naturalist, keenly interested in zoology, botany and horticulture. Basing their study on the *Tuzuk-i-Jahangiri* and court paintings, particularly the portraits of animals and flowers by Ustad Mansur, they observe that the *Tuzuk* provides data on natural history and science with extreme accuracy, meticulousness and originality, giving descriptions of not only ecology and habitats but also weights and measures. Demonstrating Jahangir's trenchant desire to research the fauna, Alvi and Rahman mention the emperor's contribution to the study of the sarus crane and the gestation periods of elephants.[103] Ardent about clean environment, Jahangir

[101] 'The Moghul Emperors of India as Naturalists and Sportsmen' included in the volume as chapter 1.

[102] Shibani Bose in her article 'From Eminence to Near Extinction: The Journey of the Great One-Horned Rhino', in Rangarajan and Sivaramakrishnan, eds, *Shifting Ground*, pp. 65–87 demonstrates the interconnectedness of the fate of the rhinoceros with changing human cultures and settlement patterns till into the first millennium CE. With the greater utility of elephants and horses in the war in the second millennium, the rhinoceros lost its importance though it was found in central and north India, even if in somewhat diminished numbers. The value of the rhinoceros declined with the introduction of modern weaponry in the nineteenth century. Its horn or other parts of the body were no longer fancied for its medicinal properties or as a delicacy in food. Sport rather than sustenance or distinction combined with lethal weapons confined the rhinoceros to its limited habitat in the twentieth century. Also see Rangarajan and Sivaramakrishnan, 'Introduction', pp. 10–12 in the same book.

[103] Divyabhanusinh complements the findings of Alvi and Rahman by adding that Jahangir also studied the gestation period of the cheetah and calculated it as over 75 days. However, considering the 12 instances of

ordered experiments to test the compatibility of air of Ahmadabad and Mahmudabad to human health. Fascinated by the sciences of healing and medicine, Jahangir asseverates the healing powers of *mumiya* (bitumen) and opium for human beings and cheetahs respectively.[104] Different in orientation from the two papers cited above, there are several other scholarly works that have studied the interaction between man and nature during the Mughal period. One such valuable contribution is that of Shireen Moosvi[105] who has challenged the notion of a pristine, harmonious pre-colonial environment and argued that the survey of the natural environment of human society in India—whether the domain of the cultivated lands, forests, forest products, wild life and river courses—show vast changes that had taken place during the last 400 years. Talking about man's natural environment and the relationship between man and nature, she emphasises on the inter-connectedness and the mutuality between different environmental elements during the sixteenth and seventeenth centuries. Her analysis of economic geography and natural environment clearly demonstrate the idea of agrarian environment. Agricultural zones, she observes, were interspersed with sizeable thick and impenetrable woods that not only had ecological significance, but were also an important economic factor and a necessary requirement for agriculture. They provided pastures for cattle, and space for nurturing firewood, timber and forest products like wild fruits, honey and toddy. The forests also served as a refuge for the zamindars to escape the heavy revenue collection by the Mughal state, the extent and type of which (forests) during this period can be inferred from the locations of hunting grounds and/or the habitats of wild animals, for example, while the presence of the wild elephants would indicate the proximity of a dense forest, that of the wild cheetahs would imply the existence of grasslands

---

cheetahs from Africa, breeding between 1964 and 1968, Divyabhanusinh argues that the period of gestation for the cheetah varied from 86 to 95 days. See Divyabhanusinh, *The End of a Trail, The Cheetah in India*, Third Edition, New Delhi: Oxford University Press, 2006, p. 42.

[104] Compare Divyabhanusinh who talks of the use of *tiryaq-i-faruqi* (preparation of opium) to heal the antelopes. *Ibid*.

[105] Shireen Moosvi, 'Man and Nature in Mughal Era', Symposia Paper 5, Thematic Symposium 'Man and Environment in Indian History', *Proceedings of the Indian History Congress*, 54th Session, Mysore, 1993.

and scrubs in the vicinity. All these, Moosvi has argued in this empirically rich paper, could be again ascertained by examining the statistics of area under cultivation for the different regions.

The importance of nature, the intertexture between different modes of livelihood and human–environment interaction is visibly exemplified by the papers included in the section on 'Forests, Pastoralism, Nomadism'. Chetan Singh raises the issue of the significance of environmental factors in determining the economic development of the Mughal Empire and understanding the evolution of state and society in pre-colonial South Asia in the article 'Forests, Pastoralists and Agrarian Society in Mughal India'[106]. Reiterating that different aspects of environment cannot be studied in isolation neither can agriculture be seen exclusive of the other elements of environment, though it has long been assumed that the prosperity of the Mughal state depended on its agrarian base, Singh argues that it was not, in fact, upon the settled villages and the revenue-yielding lands that the stability of the Mughal society rested; there was much that lay beyond the agrarian economy. The Mughal ruling class did not possess absolute monopoly over geographical space and/or natural resources even if they fell within the political confines of the empire. Ecological diversity—a mix of production, forests and grasslands—prevailed in the empire. This fostered a variety of socio-economic systems that necessitated a relationship of sharing with other social organisations that used the environment in diverse ways. The extent of forested or uncultivated territory was considerable with several *qasbas* (rural market towns), towns and adjoining agricultural fields in the midst of the forests. While many tribes were fairly dependent on pastoralism, they were also engaged in economic exchange with the adjoining agrarian society and were added as new social groups to the sedentarised agricultural population. The lives of the peasants and pastoralists, Singh avers, were inextricably entwined. Their involvement in the mutual exchange of products was accompanied simultaneously with the scramble for encroachment on each other's space: agriculturists treaded on the grazing grounds of the pastoralists and the herdsmen who wandered on the fringes of productive areas re-appropriated the abandoned fields and habitations.

[106] Included in this volume as chapter 3.

Inquiring further the tribal subsistence patterns, Archana Prasad[107] studies the ways in which the survival practices and livelihood patterns of the forest communities were related to their local and regional agrarian economies. Arguing in the context of the Baiga of Eastern Central Provinces, for whom the forests were a primary source of survival and who believed that the forests had all such elements that fulfilled their living requirements, she challenges the notion of isolation of forest communities and their conflicting interests with the local economy and demonstrates that the stability of the subsistence economies of tribal groups was embedded in multiple seasonal strategies, related to one another through semi-sedentary lifestyles or patterns that required seasonal mobility from one form of production to another. The Baiga livelihood depended on manifold occupational structure: a combination of gathering, hunting and *bewar* (a Baiga term for swidden cultivation). Of these, bewar was considered primary and privileged in Baiga mythology that enhanced the social status of the Baigas *vis-à-vis* the other forest communities; moreover, their ritual role during sowing and harvesting reinforced their image as the kings of the forests. Gathering and hunting were determined by the bewar agricultural cycle and were undertaken when bewar was not possible. The three survival strategies of the Baigas bound by fragile seasonal rhythm could succeed only in a forest with diverse forest species and entailed local connections with peasants and artisans who visited the forests and used forest produce; an assessment of the Baiga activities in forests indicate that the Baiga economy was closely linked with that of the peasants, graziers and craftsmen who were important users of the forests.

Presenting yet another aspect of pastoral and nomadic communities, Bhangya Bhukya (chapter 5) discusses the process of subalternisation of the lambadas, a pastoral, nomadic community engaged in caravan trade, particularly in the southern part of India. Communities of nomadic caravanners were found all over the subcontinent, each with its own distinctive historical, cultural and social background. They were loose groupings, assimilating members from other groups or from settled people who were involved in such a trade.[108] The lambadas had

[107] 'The Baiga: Survival Strategies and Local Economy in the Central Provinces' (included in this volume as chapter 4).

[108] The caravan trade dates back to ancient times. From the eleventh century, as trade and commerce expanded, caravan trade began to flourish

emerged as prominent transporters since the Delhi Sultanate. They accompanied the armies of the Delhi Sultans, Mughal emperors, Nizam of Hyderabad and the English East India Company as independent transporters and also served as commissariat foodgrain and baggage transporters to armies during wars. Their caravan trade networks were highly organised and played a crucial role in the development of distance trade from the sixteenth century. Their contribution as caravanners was particularly significant during famines since they supplied foodgrain from surplus to deficit areas through their caravan network. Though they served the other merchants and dealers as carriers, they themselves emerged as merchants, a fact that suggests that at least some had accumulated capital. However, with the rise of regulated market and new economic relations enforced by the colonial rulers, the caravan traders, argues Bhukya, were marginalised, reducing the 'freewheeling' lambada traders to the position of hired labourers. Their low status in the trade hierarchy resulted in their exploitation and denied them institutional support and access to credit, making their lives precarious.

Studying the complexities associated with interactions of various components of environment in the historical narratives of pre-colonial India, water, irrigation and such other resource strategies figure as elements that define farm environments of South Asia more than any other physical environment. Mayank Kumar, in the article 'Situating the Environment: Settlement, Irrigation and Agriculture in Pre-Colonial Rajasthan' (chapter 6), examines the availability and utility of various inherent water systems and their intricate role at mitigating the vagaries of nature. He invokes the importance of the linkages between the environment, social structure, cultural values and the urgency for the state to monitor the various issues related to the appropriation and uses of water. Focusing on the necessity of studying various systems of water management that were developed in arid and semi-arid regions with unequal distribution of rainfall and low water table that was often saline, he argues, in the context of semi-arid and arid Rajasthan that the management of water available during the three months of

reaching its culmination during the Mughal period as the imperial army maintained and employed large numbers of nomadic groups as carriers and transporters of baggage and foodgrain.

monsoon was and continues to be crucial for sustenance. The general geophysical characteristics of the region of Rajasthan suggest that there were inadequate water resources leading to limited possibilities of human settlement and agriculture. This entailed several initiatives not only at the community level but also state enterprise on a large scale for irrigation and potable purposes and its active role in influencing agrarian relations and promoting agriculture. People maximised the use of rainwater by evolving several indigenous methods of water harvesting for irrigation. These interventions manifest environmental, social and political concerns and an attempt to protect and preserve sedentary settlements and production.

The discussion on the political economy of irrigation and water relations in agricultural production in pre-colonial India is taken up by Tripta Wahi in chapter 7 of this volume. Invoking the theme of irrigation, state and society in pre-colonial India, she emphasises the importance of water as a valuable analogous constituent to land in agricultural production. Defining irrigation as human intervention in water with the objective of producing bio-mass and an essential component for agricultural production as also for the structures and mechanisms of appropriation of surplus produce, she focuses on two modes of irrigation—canals constructed on all running surface water that had multi-dimensional impact on state and society evident in economic activities comprising agriculture, horticulture, trade and huge revenue collections; and wells that were field-specific with their source underground. Wells had complex proprietary rights distinct from land rights of a landowner making wells a potential site for contestation of water rights, with the ownership ranging from the individual, family, community, state or whoever facilitated the sinking of the well, though the state through various administrative means maintained a close watch over the wells and was deeply engaged in creating and controlling irrigation facilities. In fact no wells could be constructed without the permission of the state—Sultanate, Mughal or regional—for it were, as argued by Wahi, the water rights that defined the nature of the state. Though the role of the state and the society was distinct in these two modes of irrigation, there were common patterns in relation to water rights, appropriation of labour and socio-politico, economic consequences for both the state and the society.

Probing the impact of water systems on the environment and human–environment relations, I discuss the changing nature of river courses in north India in chapter 8. Focusing on the corrosive power of rivers, the destructive as well as beneficent effects of their impetuous trajectories and the social implications of this process, the emphasis is on the significant role of the water courses in determining human settlements and human activity. This also explains the abundance of archaeological sites along river margins and the banks of streams and lakes. Transformations in landscapes occur whether owing to geology, geomorphology or human activity, and while there is a need to study it, it is even more important to study the impact of changing landscapes on history. Perhaps, the most effective way, as this essay suggests is to examine the alteration in river courses and their effects. To understand the relation of river action to geographic forms and how on some occasions, the changing course of rivers caused feuds over land rights, the analysis is based on a case study of one of the major river systems in South Asia, river Ganga (Ganges) and its tributaries and distributaries in north India. For instance, the recurrently changing course of river Gandak (tributary of Ganga) in Mughal sarkar Gorakhpur had created *gangshikisht* and *gangberamud* lands, that is, lands carried away or thrown away by the changing course of the rivers resulting in innumerable disputes over these lands. These disputes illustrate that several zamindars across sarkar Gorakhpur and particularly in regions affected by the impetuosity of river Gandak and its tributaries remained engaged in a continuously recurring discord. In the discussion of the changing course of river Ganga and its tributaries, one notices the turbulent hydrograph of the rivers which have caused natural disasters in the form of floods, tides and inundated banks. Erosion (silting up of one bank while the other is being eroded) and avulsion (the cutting of a new channel) have been its major characteristic and method of altering its course. Yet its long historical past has demonstrated its enduring nurturing qualities, its self-regulating mechanism of recuperating and reviving the areas it had inflicted catastrophe upon.

The holistic view of the environment and its issues entail the understanding of the relationship not only between the people and the lands but also the complex encounters of humans with animals that help to explore the attitude towards the animals and study the fauna in their own spaces. This may facilitate to determine whether historical processes had any impact on the spatial distribution of the animals and whether it contributed to the extinction of certain species. Several scholarly works have contributed to the understanding of the history of the animal, man–animal interaction, fauna in their own worlds, the gradual erosion of their spaces and the extinction of some species.[109] Simon Digby (chapter 9), however, focuses on a different kind of relationship with the animals and its significance when he discusses the importance of elephants in medieval Indian warfare.[110] He suggests that the main source of war-elephants for the Delhi Sultans was Bengal. In fact, till recent times, the thick jungles of eastern Bengal have remained an important source of wild elephants. Commenting on the source of supply, Digby observes that the elephants could come into the Sultan's *pil-khana* (elephant stables) as plunder and/or tribute from subordinate rulers or governors, by purchase from outside territories or trapped directly from a 'wild state'. Describing the number of elephants in the pil-khana during the period of the Delhi Sultans and comparing them with those in the Ghaznavid kingdom (which showed a higher number in the pil-khana), he remarks that the north Indian breeding grounds of the elephant were probably diminished by over-exploitation or by settlement and cultivation between the early eleventh and thirteenth century, thus making the wild elephant less plentiful in northern India during the thirteenth–fourteenth centuries. Nonetheless, the actual possession of elephants was considered an important symbol of royalty, power and authority in the medieval times. In the struggle for power in the Delhi Sultanate during 1388–98 CE, the possession of a pil-khana was deemed to be one of the most important assets. It was for this reason, perhaps, that the employment of the elephants by the Mughal emperors in the sixteenth

[109] See, for instance, Divyabhanusinh, *The End of a Trail*; also his *The Story of Asia's Lion*, Mumbai: Marg Publications, 2005; also his 'Lions, Cheetahs and Others in the Mughal Landscape', in Rangarajan and Sivaramakrishnan, eds, *Shifting Ground*, pp. 88–108; also see Raman Sukumar, *Asian Elephant: Ecology and Management*, Cambridge: Cambridge University Press, 1989; also Thomas R. Trautmann, *Elephants and Kings, An Environmental History*, History Series, Ashoka University, Ranikhet: Permanent Black in association with Ashoka University, 2015.

[110] In India, the elephants were employed in warfare since fourth century BCE though they may have been domesticated some centuries before that.

century was considerably high. In fact, elephants were trapped extensively during this period in the regions south of the Gangetic plain.[111] Digby notes that while Akbar was a connoisseur of elephants and procured elephants from many places in northern India, Jahangir captured only 69 elephants in a royal hunt in Gujarat. Such efforts notwithstanding, the association with elephants witnessed a transformation by the end of the sixteenth century and possibly a few decades before this.[112] With the rapid development of firing artillery and handguns and with improved musketry in the seventeenth century, its employment in battles was rendered almost obsolete, making it principally the transport, command vehicle mounted by generals and commanders and by the nineteenth century, elephants had become a popular transport commodity to access the dense, impenetrable lands. Having lost their military role by the seventeenth century, elephants nevertheless served as 'high platforms' from where the commanding generals could watch, and be watched from the battle scene—a consequence of the shift towards warhorse economy and gunpowder weaponry.[113] The focus switched to exploiting the 'locomotive' potential of elephants, warhorses, dromedaries and oxen,[114] possibly to realise the constant urge of the Mughal emperors

[111] Compare Pratyaya Nath, 'Rethinking Early Mughal Warfare: Babur's Pitched Battles, 1499–1529', in Raziuddin Aquil and Kaushik Roy, eds, *Warfare, Religion and Society in Indian History*, New Delhi: Manohar, 2012, pp. 109–45; also his, 'From Mocking to Becoming the "Lords of the Elephants", War-Elephants and the Mughals, 1526–1590', Paper Presented at a Workshop on 'Environment and Histories' organised by Nehru Memorial Museum and Library, New Delhi, 12 May 2012.

[112] Compare Trautmann, *Elephants and Kings: An Environmental History*.

[113] See Gommans, *Mughal Warfare, Indian Frontiers and High Roads to Empire, 1500–1700*, London and New York: Routledge, 2002, pp. 99–100.

[114] Gommans argues that although horses were generally found in all parts of the subcontinent, the ecological conditions—the extensive grazing areas, suitable green and dry fodder crops—for horse breeding in India were hardly satisfactory. Under these circumstances, early British veterinary specialists considered the so-called *dub* grass the most suitable fodder for horses. The Indian alternatives for grass and hay were not very attractive either and remained almost negligible till the nineteenth century. Both oats (main fodder crop in Europe) and barley (main horse feed in Arabia) were also in short supply in India. Good breeding necessitated selecting the best stallion and mares but in India the practice was of importing strong foreign stallions to maintain

(Babur to Aurangzeb) for persistent, unremitting mobility to bridge the inner frontiers of the empire.

Asok Kumar Das and Mahesh Rangarajan[115] take us into a different sphere—that of hunting—to explicate the man–animal relationship. They comment on the ambience and the environment that facilitated the Mughal passion for hunting. Das observes that the thick jungles, vast stretches of grasslands, wooded hills and difficult terrains of India sheltered much game and wildlife. Large tracts of land were reserved as the royal preserves where none but only the king could hunt, it being an imperial prerogative. Members of the royal family, including the *zenana* (harem), the important courtiers and retainers, participated in these hunts. Specialised hunts, he argues, were tailored to the specifics of the area and its ecology, for instance, the *qamargah* or ring hunt, *shakhband* or where the area was fenced off with the branches and trunks of trees. While the Mughals were keen on falconry, fishing, crocodile shooting and elephant trapping, they specialised in hunting the deer/antelope by using trained cheetahs. Cheetah hunting, Das suggests, was probably not known before 1555 CE since there is no mention of it in the *Baburnama*, but the cheetah became an obsession with Akbar, who collected almost nine thousand cheetahs during his reign and keenly followed, supervised and participated in the trapping of the cheetah. Mughal compulsive enthusiasm and fondness for the cheetah encouraged its depiction in folk tales, court paintings, book covers, textiles, carpets, metal ware and also in musical melodies or the *ragamala*.[116] While the

---

the quality of the indigenous breed. Despite these drawbacks, horses were considered an effective instrument of power, a token of royalty, nobility and chivalry by the Mughal emperors. The dromedaries, on the other hand, were highly sensitive to the vicissitudes of climate and could survive in the relatively flat and dry terrain; the ox, an in-bred species, could survive in both dry and humid conditions. While the ox travelled slowly and could carry only about a third of the dromedary's burden, its advantage was its all-India reach and its military power and capacity to pull the cannons. Gommans, *Ibid*., pp. 111–30.

[115] The articles of both are included in this volume. The original article of Asok Kumar Das contains seven marvellous paintings depicting the cheetah in different hunting scenes, which unfortunately could not be reproduced in this volume for reasons of copyright.

[116] Also see Divyabhanusinh, 'At The Court of The Great Mughals', *The End of a Trail*, pp. 36–77.

cheetah remained prominent during the Mughal period, it was from then on, Das asserts, that the population of the cheetah dwindled rapidly and by the 1950s, this swift and astounding wild cat became extinct in the Indian subcontinent.

Connecting hunting with the Mughal landscape, Mahesh Rangarajan observes that the culturally and ecologically diverse Mughal-ruled territory was a boon to the Mughal ardour for hunting. While agriculture was the core of the Mughal economy and the peasants undertook continual reclamation of land to facilitate cultivation, large areas remained uncultivated and forested, with some under dense jungles. Such patterns nourished the Mughal zeal and fervour for hunting. Hunting during this period, argues Rangarajan, acquired a sense of pleasure and relaxation; it became a form of physical exercise for militarised nobility and cavalry, assumed the character of a ritualised activity with a political connotation and a way to camouflage armed expeditions, especially against rebels. Hunting endured as a royal privilege during the period though the hunting grounds of the Mughal emperor were not completely out of bounds for the common people.[117] He adds that the environment aided both sedentarisation and hunting but it was the former that prevailed and overwhelmed affecting adversely the habitats of large mammals like the elephant, rhino, tiger and the lion, heralding changes in the landscape.

The aforesaid discussion is proposed as a preface to several thoughts and ideas pertaining to environment studies, to understand and evaluate human–environment interaction and human-induced changes, the scale of which has been historic with tremendous implications for the modern ecologies. Yet the aspect of transition cannot be considered new. There was nothing primeval or pristine about the pre-colonial past. Neither can the medieval–early modern period be seen as a mere criterion to study and ascertain the environmental processes in the subsequent centuries. In fact, no study of the environment can be seen as fulfilled unless studied in the larger, broader perspective.

[117] Barbara N. Ramusack has discussed colonial hunting practices and their appropriation and reinterpretation by Indian princes. See her *The Indian Princes and their States*, first Asian edition, New York: Cambridge University Press, 2005; also John M. Mackenzie, *The Empire of Nature: Hunting, Conservation and British Imperialism*, New York: Manchester University Press, 1988.

SECTION I

NATURE AND ITS PERCEPTION

The Moghul Emperors of India as Naturalists and Sportsmen*

Salim A. Ali

The title of this chapter is somewhat misleading; hence it may be advisable at the outset to indicate the scope. The term 'Moghul Emperors' here represents only the Big Six, from Babur—the illustrious founder of the dynasty—to Aurangzeb, with whose death the great empire launched on a career of steady and rapid decline.

The 'Naturalists' of the title also needs qualification. It stands here only in respect of animal life, though it is well known that the Moghuls were great lovers of nature in all its other aspects as well.

The wonderful gardens[1] built by them all over northern India remain to this day to bear testimony to their love for flowers and trees, and the genuine delight which Babur and his great-grandson Jahangir felt in the natural objects they saw around them, cannot help impressing anyone who wades through the inimitable memoirs left by these two sovereigns.

Babur

Babur was a born commander and leader of men, possessing all the attributes that go to make a successful general, conqueror and administrator, and withal a very loveable and extremely human man.

He was an athlete of the highest order and a sportsman in every sense of the word. His nature was aesthetic to a degree, and all throughout the delightful memoirs written by him, one constantly alights on passages which reveal something of the passionate infatuation he felt for

* Salim A. Ali, 'The Moghul Emperors of India as Naturalists and Sportsmen', *Journal of the Bombay Natural History Society*, volume xxxi, issue number 4, 1927, Part I, pp. 833–45.

[1] For descriptions of the gardens, see C. M. Villiers Stuart, *Gardens of the Great Moguls*, London: Adam and Charles Black, 1913.

the beautiful and the new, whether in scenery or architecture, plants, flowers or animals.

One of Babur's first cares after his victory at Panipat in 1526 was to describe at length the land of his acquisition, its peoples, customs, animals, fruits and flowers: to compile, in fact, a comprehensive Gazetteer of Hindustan. The outstanding feature of his accounts is their extreme truthfulness and accuracy. If there is anything of which he is not positive at the time of writing, he does not omit to make special mention of the fact, and hearsay of the veracity of which he is not convinced is likewise duly recorded as such. For instance, writing about parrots he says, 'I had imagined that a parrot or *sharak* (*Myna*) only repeated what it had been taught, and could reduce nothing into words from its own reflections. Abul Kasim Jalair who is one of my most familiar servants lately told me a remarkable incident. A cage of a parrot of this last-mentioned species[2] having been covered up, the parrot called out, "Uncover my face, I cannot breathe." On another occasion the bearers who were employed to carry it had set it down to rest themselves and a number of people passed by, the parrot called out, "Everybody is going by, why don't you go on?" Let the credit rest with the relater! Yet till one hears such things with his own ears he never can believe them.'

Similarly, in another place, after describing the *Lujeh* (monal pheasant), he says with the keenest humour, 'A remarkable circumstance is told of them. It is said that in winter they come down to the skirts of the hills and if in their flight one of them happens to pass over a vineyard, he can no longer fly and is taken. God knows the truth! Its flesh is very savoury.'

Whether engaged in weighty affairs of state, or in marching against a foe, Babur always had his senses wide awake to objects around him beyond the pale of his immediate concern. A new flower or bird or beast never failed to excite in him a feeling of the profoundest interest. He made a careful mental note of the object, and reproduced it faithfully in his memoirs whenever he found a temporary respite from the arduous duties of kingship in a newly conquered and unsettled country. For example, the *Pied Myna* (*Sturnopastor contra*) finds a place in his memoirs thus, 'When I threw a bridge over the Ganges and crossed it, driving

[2] The large Indian Paroquot (*Paleornis nepalensis*).

the enemy before me, I saw in Lucknow, Oudh and these countries, a species of sharak which has a white breast and a piebald head with a black back. I had never seen it before. This species probably does not learn to speak at all.'

Of the larger mammals of Hindustan, the rhinoceros was one that must have seemed altogether strange and unnatural to the conquerors, and Babur took great delight in hunting the beast. In his memoirs he gives the following account of a hunt. (This was at the time of his final march against Hindustan which ended successfully at Panipat):

> We continued our march till we came near Bekram (Peshawar) and then halted. Next morning we continued halting in the same station, and I went out to hunt the Rhinoceros. We crossed the Siah-ab (that is, Black River—perhaps another name for the Bara) in front of Bekram, and formed our ring lower down the river. When we had gone a short way, a man came after us with notice that a rhinoceros had entered a little wood near Bekram and that they had surrounded the wood and were waiting for us. We immediately proceeded towards the wood at full gallop and cast a ring round it. Instantly on our raising the shout the rhinoceros issued out into the plain. Humayun and those who had come from the same quarter (that is, from Turkestan) never having seen a rhinoceros before, were greatly amused. They followed it for nearly a kos, shot many arrows at it and finally brought it down. The rhinoceros did not make a good set at any person or any horse. They afterwards killed another rhinoceros. I had often amused myself by conjecturing how an elephant and rhinoceros would behave if brought to face each other; on this occasion the elephant-keepers brought out the elephants so that one elephant fell right in with the rhinoceros. As soon as the drivers put their beasts in motion, the rhinoceros would not come up but immediately ran off in another direction.[3]

[3] Captain Williamson, in his *Oriental Field Sports*, London: Young, 1819, recounts several incidents of the deadly enmity that is supposed to exist between the rhinoceros and the elephant and the stories copied from this not over-veracious source have subsequently found their way into European works on Natural History. Despite its bulk and strength, the rhino is, as a rule, quiet and inoffensive. Owing to the nature of the terrain they inhabit, elephants are almost always used in rhino hunting in India. As a rule, the mounts remain indifferent in the presence of the enemy though occasionally, one takes fright and bolts.

Further on Babur mentions, 'In the course of my expeditions into Hindustan, in the jungles of Peshawar and Hashnagar I frequently killed the rhinoceros. It strikes powerfully with its horn, with which in the course of these hunts, many men and horses were gored. In one hunt it tossed with its horn a full spear's length, a young man named Maksud, whence he got the name of "Rhinoceros Maksud" (i.e. Rhino's aim).'

On the conclusion of Babur's accounts of the fauna of Hindustan, descriptions of the trees, flowers and fruits follow, and throughout is noticeable the undercurrent of his superlatively aesthetic soul, and the knowledge and insight born of careful and intelligent observation. It appears truly remarkable that a man in Babur's situation, faced with innumerable and ever-recurring practical and administrative difficulties in a foreign and unsettled land, with the prospect of rebellion within and invasion without perpetually staring him in the face—with marked discontent amongst his troops and following to boot—should have found time to devote himself so earnestly to minor pleasures which would probably have had no appeal for lesser natures than his own.

Humayun

On the death of Babur, his son Humayun ascended the throne of Hindustan. Unfortunately, he appears to have kept no memoirs of himself and the chief contemporary records available to us are the *Tezkereh-al Vakiat*[4] (or private memoirs of the emperor Humayun) written by his confidential domestic Jouher, and the delightful *Humayun Nama*[5] of his sister Gulbadan Banu Begum.

These tell us extremely little of the aspect of Humayun's nature that concerns our chapter, but however suffice to establish the fact that he had inherited in full measure his father's love of nature and fondness for sport.

That our information as regards Humayun's private life and affairs should be so scanty is not surprising for throughout his reign, he was

[4] Jouher, *The Tezkereh-al Vakiat* or *Private Memoirs of the Moghul Emperor Humayun*, trans. Major Charles Stewart of the Hon'ble East India Company's Service, London: Oriental Translation Fund of Great Britain and Ireland, 1832.

[5] Gulbadan Begum, *Humayun Nama*, trans. Annette S. Beveridge, London: Royal Asiatic Society, 1902.

being unremittingly harassed by his rebellious brothers, and it was not long after his accession that he was even driven out of his kingdom and was not able to return and regain it till three or four years later.

No elaborate accounts of Humayun's hunts are available. That he showed keen partiality for and considerable skill in the chase, however, may be gleaned from several references in the memoirs of his father. With paternal joy and pride, we are told how on one occasion, when the prince was 11 and Babur and several of his associates were in a boat in the *Bagh-e-banafsheh* (Garden of Violets) in Kabul, Humayun 'shot a waterfowl in very handsome style.'[6]

Further on is recounted[7] how the sight of a rhinoceros—an animal Humayun had never seen before—at Peshawar had amused him and with what enthusiasm he and his party had followed the beast and brought it down and then had killed another. Hunting an animal like the rhino with bows and arrows, which were the weapons employed on this occasion, is a feat that calls for not only skill and physical strength but an uncommon amount of that dash and courage which were ever characteristic of the great Moghuls. And this was at a time when the prince was barely 17.

That Humayun was naturalist at heart is evident from just this one little passage that appears in the *Tezkereh*. To realise the true significance of the incident, it must be remembered that it happened at a time when, defeated by Sher Shah and deprived of his kingdom, he was fleeing for his life and liberty through the inhospitable desert of Sind—a harassed refugee—subjected to the greatest hardships from hunger and thirst, and accompanied only by a mere handful of his faithful adherents. At Amarkot, '… the king undressed and ordered his clothes to be washed, and in the meantime wore his dressing gown. While thus sitting, a beautiful bird flew into the tent the doors of which were immediately closed and the bird caught. His Majesty then took a pair of scissors and cut some of the feathers off the animal; he then sent for a painter and had a picture taken of the bird and afterwards ordered it to be released.'

[6] Babur, *Baburnama, Memoirs of Zahiru'd-din Muhammad Babur, Emperor of Hindustan*, translated from Chaghatai Turki by John Leyden and William Erskine in two volumes, annotated and revised by Sir Lucas White King, London and New York: Humphrey Milford, Oxford University Press, 1921, vol. II, p. 138.

[7] *Ibid.*, p. 159.

A temperament capable of being roused from the gravest anxiety and concern to light-hearted pleasure and interest at the mere sight of a strange and insignificant bird surely proves a more than ordinarily deep-rooted love for nature. In spite of the fact that Humayun's reign was short and replete with trouble and anxiety, he sometimes found leisure for indulging in the chase, for which sport as a carefree prince we know he had evinced marked predilection. His sister tells us[8] that one day at Kabul, 'His Majesty attended by Mirza Hindal (his brother) was hunting near the mountain passes. They had very good sport. The Emperor went to where the Mirza was hunting and had made a very good bag. Following the rules of Chingiz Khan the Mirza proffered his game to the Emperor, for it is the rule of Chingiz Khan that inferiors should so act towards their superiors. In short he gave the Emperor all his game…'

This account also serves in bringing to light the peculiar hunting etiquette of the Moghuls, traces of which are to be met in our own day, especially in relation to the hunts of Rajas, Nawabs and other high personages.

It is much to be regretted that Humayun's career as a naturalist was so disturbed and short. He was a devout lover of nature, and given the opportunity, we should certainly have expected to be left with notes and descriptions at least equalling in originality and interest to those of his father or of his grandson Jahangir.

Akbar

Akbar, rightly called 'the Great' reigned over Hindustan from 1556 to 1605. Summing up his character, Dr Richard von Garbe[9] states, 'Akbar was very fond of flowers and perfumes and especially enjoyed blooded doves (pigeons) whose care he well understood. About 20,000 of these peaceful birds are said to have made their homes in the battlements of his palace.' His historian (Abu'l Fazl) relates, 'His Majesty deigned to

[8] Gulbadan Begum, *Humayun Nama*, pp. 96–97.

[9] Richard von Garbe, *Akbar, Emperor of India, A Picture of Life and Customs from the Sixteenth Century*, trans. Lydia G. Robinson, Chicago: The Open Court Publishing Company, 1909.

improve them in a marvellous manner by crossing the races which had not been done formerly.'

Akbar was passionately fond of hunting and pursued the noble sport in its different forms, especially the tiger hunt and the trapping of wild elephants, but he also hunted with trained falcons and leopards. He was not fond of *battue* (a hunt in which hunters force the game to flee in the direction of the hunter); he enjoyed the excitement and exertion of the actual hunt as a means of exercise and recreation for training the eye and quickening the blood. Besides chess, cards and other games, fights between animals may be specially mentioned, of which elephant fights were the most common but there were also contests between camels, buffaloes, cocks and even frogs, sparrows and spiders.

In support of his conclusion that Akbar was true to the traditions of his ancestry and a brave man, von Garbe quotes the following incident, 'On the way back to Agra where at the time he was holding court, Akbar had ridden alone in advance of his escort and suddenly found himself face to face with a powerful tigress who with her five cubs came out of the shrubbery across his path. His approaching attendants found the nineteen-year-old emperor standing quietly by the side of the slaughtered beast whom he had struck to the ground with a single blow of his sword. To how much bodily strength, intrepidity, cold-blooded courage and sure sightedness this blow of the sword testified, which dared not come the fraction of a second too late, may be judged by any one who has any conception of the spring of a raging tigress anxious for the welfare of her young.'

Two other incidents are recorded[10] where the emperor saved the life of a man who was in the act of being mauled by a tiger by shooting the animal dead when his following were in a state of panic and complete disintegration, and incapable of coming to the victims' assistance. Other equally outstanding examples of his cool-headed courage and daring are not wanting, and some of Abu'l Fazl's descriptions of the fights of infuriated and *musth* elephants, one of them being ridden and guided by young Akbar, though couched in such flowery and ornate style as to render the incidents somewhat melodramatic, are sufficient to convince the most exacting of his wonderful prowess, pluck and bravery.

[10] Abu'l Fazl 'Allami, *A'in-i Akbari*, trans. H. Blochmann, Calcutta: Asiatic Society of Bengal, 1927, vol. I, p. 284.

Akbar was passionately fond of animals, and the royal menagerie was a very extensive one comprising, as Abu'l Fazl states, 'Animals of all kinds from Persia, Turkestan and Kashmir, whether game or other' which had been brought together 'to the wonderment of beholders.' Bernier[11] mentions that the inmates of this menagerie were led past under the royal window where the monarch sat every day about noon, the procession commencing with the horses and elephants. 'Other animals are next introduced' continues the writer, 'tame antelopes kept for the purpose of fighting with each other;[12] Nilgai or grey oxen that appear to me to be a species of elk; rhinoceroses; large Bengal buffaloes with prodigious horns that enable them to contend against lions and tigers; tame leopards or panthers employed in hunting antelopes; some of the fine sporting dogs from Usbec of every kind, and each dog with a small red covering; lastly every species of birds of prey used in field sports for catching partridges, cranes, hares and even it is said, for hunting antelopes on which they pounce with violence beating their heads and blinding them with their wings and claws.'

Abu'l Fazl informs us that Akbar paid great attention to the selection and breeding of elephants, camels, cows, mules and horses and that through his encouragement of the last, the breed of horses produced in Hindustan was as fine as those of Irak (Iraq) or Arabia.

One of the modes of hunting most frequently employed by the Moghul emperors was the qamargah or 'ringing-in' method. Large tracts of country were surrounded by the armies which gradually worked their way towards the centre driving in and collecting the game. On account of the area over which operations extended, very often hundreds of square miles, the processes occupied several months. The various divisions of the army were placed under the command of their proper officers, and the whole thing was really in the nature of what are now known as Army Manoeuvres. As the favourite plan of campaign of the Moghuls was to surround an enemy first and then

[11] Francois Bernier, *Travels in the Mogul Empire, A.D. 1656–1668*, trans. Archibald Constable, London: Oxford University Press, 1891.

[12] Emperor Akbar was particularly fond of this sport and in the *Ain.*, pp. 218–22 are to be found full details regarding the kinds of fighting deer, how they fought together, and elaborate regulations as to the betting allowed on such encounters.

gradually close in upon him, these qamargah hunts provided the means of keeping the soldiery in the necessary training and practice in times of peace.

One such 'drive' that took place under the orders of emperor Akbar in the year 1567 is remarkable for the magnitude of the scale on which operations were carried out. On this occasion, 50,000 beaters were employed and, according to Abu'l Fazl, involved all the country 'from near the mountains on the one side and from the River Bihat (Jhelum) on the other.' The historian proceeds, 'Each district was made over to one of the great officers and Bakhshis, Tawacis and Sazawals were appointed to every quarter. Several thousand footmen from the towns and villages of Lahore Province were appointed to drive the game. A wide space within ten miles of Lahore was chosen for the collecting of the animals.'

This drive occupied a whole month. When all the arrangements were completed, 'His Majesty the Shah (Akbar) went to the hunting ground and viewed it from the circumference to the centre. Every one of the Grandees and other servants who had exerted themselves in this delightful service was gratified by H. M.'s approbation. Then he placed the foot of dominion in the stirrup of auspiciousness and made his tiger-like steed career in pursuit of the prancing deer. He used the arrow, the sword, the lance and the musket. At the beginning the hunting ground was ten miles in circumference, but day by day the *Qamargah* was pushed on and its area lessened....'

Akbar had a special Game Department and caused an account to be kept of all the animals hunted with measurements and the most minute details concerning them. Particulars of the guns or other weapons used upon different occasions were also recorded. He knew his favourite guns by names and was especially fond of one which he called 'Sangram' and which afterwards came into the possession of his son Jahangir who likewise prized it highly. Akbar had a remarkable shot with this piece, and Jahangir states in his memoirs, 'He (Akbar) had no rival in shooting with a gun, and with the one with which he killed Jitmall (the defender of Chitor), and which he called "Sangram" he killed some 3,000 or 4,000 birds and beasts.' Abu'l Fazl is more moderate; he says[13] that Akbar killed 1,019 animals with 'Sangram'.

[13] *Ain*, vol. 1, p. 116.

Besides shooting with the gun, Akbar also did a good deal of hunting with the bow and arrow, and several instances are on record of his tackling with these weapons.

He was possessed of remarkable observational powers and it is said of him that he could at once tell by seeing the hide, to what hunting ground a particular deer belonged.[14]

Besides the cheetahs (it is asserted to the number of 9,000) and lynxes which largely constituted his hunting establishment, Akbar was extremely fond of good hunting dogs, and imported them from all countries. 'Excellent hunting dogs come from Kabul,' says Abu'l Fazl, 'especially from the Hazara District (north of Rawalpindi). These dogs will attack every kind of animals, and more remarkable still, they will attack a tiger.'

European bloodhounds were also imported by the Portuguese, which helped them greatly to maintain favour at court.

With regard to the birds employed in the chase, the historian says, 'H.M. is very fond of these remarkable animals and often uses them for hunting purposes. Though he trains the baz (*Astur palumbarius*), shahin (*Falco peregrinator*), shunqar (either *Falco cherrug* or *F. milvipes*) and burqat falcons (probably the golden eagle—*Aquila chrysaetos*) and makes them perform wonderful deeds, H.M. prefers the bashah (Sparrow Hawk—*Accipiter nisus*), to which class of hawks he gives various names.'

Among the various birds trained for the chase are mentioned the crow, sparrow and quail. *Odhpapers* which were brought from Kashmir, appear from the description to be some species of kingfisher. They are described as being of a blue or green colour (*sabz*) smaller than a parrot; with a red beak straight and long; and a tail rather elongated. They were taught to bring down small birds and return to the hands of their keepers.

Jahangir

If Akbar was the greatest monarch of the Moghul dynasty, it cannot be denied that Jahangir was far and away its greatest naturalist. His profuse and engrossing memoirs are a veritable natural history of the

[14] *Ibid.*, p. 290.

animals that came under his notice, and a record of the most searching observations concerning them.

It has been rightly said of Jahangir that had he been head of a Natural History Museum he would have been a better and happier man. Besides a passion for justice, the outstanding features of his character were his love of nature and his powers of observation.

Jahangir's love of the chase was excessive. He never failed to create opportunities for indulging this propensity and as a marksman showed considerable skill. He writes of himself, 'I am myself not without some skill in the use of this weapon (meaning the famous "Sangram" also called "Droostandaz" or "Straight thrower") being exceedingly fond of field sports of every kind and having frequently with the same piece killed twenty antelope in a day.' He was adept at the use of the bow and arrow, and often used these weapons especially in qamargah hunts.

Extensive game preserves were maintained and frequently the emperor hunted accompanied by the ladies of his zenana. His beautiful and accomplished queen Nur Jahan was his constant companion on such occasions. She was an excellent horse rider and possessed remarkable dexterity in handling a gun. Jahangir gives the following account of a hunt where Nur Jahan killed four tigers in quick succession. '... the huntsmen marked down four tigers and I went out to hunt them with my ladies. When the tigers came in sight Nur Jehan Begum submitted that if I would order her she herself would kill the tigers with her gun. I said "Let it be so." She shot two tigers with one shot each and knocked over the two others with four shots. In the twinkling of an eye she deprived of life the bodies of these four tigers. Until now such shooting was never seen that from the top of an elephant and inside of a howdah, six shots should be made and not one miss, so that the four beasts found no opportunity to spring or move. As a reward for this good shooting I gave her a pair of bracelets of diamonds worth one hundred thousand rupees and scattered 1,000 *ashrafis* (gold mohurs) over her.'

Like his father, Jahangir also caused minute records to be kept of his hunts with particulars as to the bag, etc. The registers showed that from the 12th year (1580) of his age to his 50th lunar or 48th solar year, 28,532 animals had been taken in his presence, including 17,167 which had been killed by the emperor himself. These are tabulated thus:

Tigers (and lions)	...	...	...	86
Bears, leopards, foxes, otters (*udbilao*) and hyenas	...	...	...	9
Blue bulls	...	...	...	889
Mhaka[15]	...	...	...	35
Black buck, chinkara, cheetal, mountain goats, etc.	...	...	...	1,670
Rams (*quj*) and red deer	...	...	...	215
Wolves[16]	...	...	...	64
Wild buffaloes	...	...	...	36
Pigs	...	...	...	90
Rang (ibex)	...	...	...	26
Mountain sheep	...	...	...	22
Arghali	...	...	...	32
Wild asses	...	...	...	6
Hares	...	...	...	23
		Total	...	3,203

Of the 13,954 birds that constituted the total bag during the period were:

Pigeons	...	...	...	10,348
Lagar-jhagar (a species of hawk)	...	...	...	3
Eagles	...	...	...	2
Qaliwaj (kites)	...	...	...	23
Owls (*chugd*)	...	...	...	39
Qautan (goldfinch?)	...	...	...	12
Mush-khwar, 'rat eaters' (probably harriers, etc.)	...	...	...	5

[15] It has been suggested that this may be the *Maha* or Swamp Deer of the Terai. As there is no mention of Sambur elsewhere in the list, it is possible these were included here. Jahangir mentions that in size this animal was equal to the Nilgai.

[16] In Persian *Kurg* is rhinoceros and *Gurg* is wolf. I think it very probable that if it is not 'Kurg' in the original MS there is at least some confusion on this point. Jahangir records the killing of a rhinoceros with a single shot in the temple, and this does not appear in the above list of game.

Sparrows	...	...	...	41
Doves	...	...	...	25
Owls (*bum*)	...	...	...	30
Ducks, geese, cranes and wildfowl	...	...	...	150
Crows	...	...	...	3,276
		Total	...	13,954
Crocodiles	...	...	...	10

Jahangir also was exceedingly fond of good hunting dogs and collected them from distant parts. Sir Thomas Roe records that once the emperor mentioned to him 'I only desire you to help me to a horse of the greatest size, and a male and female of mastiffes and the tall Irish Greyhounds and such other dogges as hunt in your lands.'

Long accounts appear in the *Memoirs* of the various hunting exploits of the emperor which are too profuse to reproduce here.

Shah Jahan

Shah Jahan ruled over the Moghul Empire from 1627 to 1665. Gifted with the love of nature and artistic temperament of his distinguished ancestors, his name is however best associated with architecture. The palace-fort at Delhi and the peerless Taj Mahal at Agra are living tributes to his magnificent regime.

Compared with his father, Shah Jahan's fondness for sport was moderate. He preferred hawking and hunting with cheetahs to the actual shooting of game.

Jahangir relates that it was once reported to him while encamped in the neighbourhood of Ajmer that there was a man-eating tiger, who had already accounted for several lives. The Prince, Shah Jahan, was detailed to 'save the people from its wickedness' and before nightfall the animal was shot and brought to the emperor.

The following is an account[17] of one of Shah Jahan's tiger hunts written by Manucci, who lived at his court for a number of years:

[17] Niccolao Manucci, *Storia do Mogor* or *Mogul India, 1653–1708*, vol. 1, trans. William Irvine, London: John Murray, 1907, p. 191.

His ordinary amusement was tiger hunting, for which he kept ferocious buffaloes with very big horns. These fought with each other or with tigers, and they are very brave animals, and skilful in the sport above referred to.

When the king desires to go out hunting the huntsmen are warned. These men see to the finding of the tigers and send out into the jungle asses, cows, sheep and goats to prevent the tigers from changing their haunts. The king goes out on his tallest elephant and the other princes likewise on elephants acquainted with the requirement of this sort of fight. They sit in uncovered howdahs, each one with his matchlock. Then they encircle the jungle with high nets, leaving only one opening, through which the king and huntsmen enter. Around the net on the outside stand a number of soldiers, who cannot wound the tiger when it comes near the net, nor can the tiger injured them, for in no manner can it break the net and get out. The order in which the king moves is as follows: In front go the buffaloes, sometimes more than one hundred in number, all in a row. On each one is mounted a man with his legs guarded by leather, and having a broad sword in one hand and holding with the other the reins, which are passed through the buffaloes' nostrils. Behind them comes the king on an elephant, and after the king the princes and the men in highest favour. When they get into the jungle where the tigers are, the buffaloes advance slowly in the formation of a half moon, until the tigers are in sight. After locating the tigers by sight and smell, a circle is formed, leaving them in the centre. In this way, the tigers finding themselves caught, search for an exit. Unable to get away, each one makes its spring in the direction that it sees best. When this spring takes place the man who is mounted on top jumps off with agility, and the buffaloes seize the tigers on their horns with great dexterity and, shaking their heads tear them to pieces. If any one of the tigers escapes the horns or refuses to stir from its place, the king fires his gun and kills it, or gives an order to kill it.

Sometimes they go out to these hunts without taking any buffaloes, but riding on elephants as I have before said. This way of hunting has much more risk for the hunters. Once it happened to King Shah Jehan that a badly wounded tiger bounded up and hung on with its claws fixed in the elephant's head. The elephant-driver fell to the ground from fright. The king seeing himself in this urgent danger, clubbed his matchlock and hit the tiger on the head with it, but the tiger did not let go, and the elephant finding he could not make use of his trunk, ran furiously till he found a tree, against which he crushed the tiger.

> It was on this account that Shah Jehan gave orders for the head of the elephants to be protected in future down to the end of the trunk with a covering of thick leather, studded with sharp nails. In addition to the huntsmen, there is always an official present whose business is to take possession of the tiger's whiskers; and therefore as soon as the tiger is dead, they put on his head a leather bag, coming down as far as the neck. Having tied the bag the officer attaches to it his seal. After this the tiger is carried in front of the entrance to the royal tents, when the official appears who has charge of the poisons, and removes the whiskers which are employed as venom.

Shah Jahan devised a novel method of punishing officials found guilty of taking bribes or of failing to discharge their duties to his subjects by getting them bitten by poisonous snakes in his presence in open court.

Aurangzeb

Aurangzeb was a man possessed of an indomitable will and courage, but on the whole appears to have taken life much more seriously than any of his forbears, and this left him little leisure for lighter pursuits.

He was fond of the chase and occasionally indulged in hunting with cheetahs and hawks. Hunting the lion was his favourite sport. The method of hunting most commonly in use even during this period was qamargah as will be seen from the following account by Bernier,[18] who was a physician at the imperial court for a number of years, 'I could never conceive how the great Mogul could hunt with an army of one hundred thousand men, but there certainly is a sense in which he may be said to hunt with 200,000 or with any number of which his army may consist. In the neighbourhoods of Agra and Delhi, along the course of the Jumna reaching to the mountains, there is a large quantity of uncultivated land covered either with copsewood or with grasses six feet high. All this land is guarded with the utmost vigilance; and excepting partridges and quails and hares, which the natives catch with nets, no person, be he who he may, is permitted to disturb the game which is consequently very abundant. Whenever the monarch is about to take the field, every game keeper near whose district the army is to pass is

[18] Francois Bernier, *Travels in the Mogul Empire.*

called upon to apprise the Grand Master of the hunt of the various sorts of game under his particular charge, and of the places where they are in the greatest plenty. Sentries are then stationed at the different roads of that district to guard the tract of ground selected, which extends sometimes four or five leagues; and while the army is on its march, on one side or the other, so as to avoid that tract, the king enters it with as many Omrahs and other persons as have the liberty to do so, and enjoys leisurely and uninterruptedly the sports of the field, varying them according to the nature of the game.'

Bernier then goes on to describe the various methods of hunting, such as with cheetahs, tiger hunts and so on.

Jahangir, the Naturalist*

M. A. Alvi & A. Rahman

Jahangir did not belong to the community of the scientists. Nevertheless, his *Memoirs* that form primarily a historical document of his reign, contain a good deal of information on natural history and science which, for its accuracy, meticulousness of observation and originality, could well qualify any author of his times to claims of such distinction. Jahangir maintained a big aviary and a menagerie, carried out observations, tests and experiments. Often he would order a specimen to be dissected in his presence, keep records for ascertaining long range phenomena and take down measurements and weights.

The painters of his court, especially *Ustad* Mansur, who held from him the title of *Nadiru'l-Asr* (unique genius of the time), excelled in the art of animal and floral portraiture. The surviving items from Jahangir's collection of the paintings make a veritable catalogue of strikingly accurate representations, which can be of use even for the purposes of modern research.

Jahangir's interests were widely known. Foreign dignitaries, his own nobles, hermits, traders and commoners brought for him presents of animals from far and wide. This added to the variety of his knowledge. He speaks of the loriquet of Malaysia, the African zebra, the mottled polecat of Afghanistan, the rare pheasants of the Himalayas, the Tibetan yak and the domestic turkey of Europe.

His intimacy with nature was far more extensive than the *Tuzuk*[1] reveals. The paintings which have survived the ravages of time show a number of birds and animals which do not find mention in the work. His establishment was full of all kinds of carnivorous beasts, hunting

* M. A. Alvi and A. Rahman, 'Introduction', in M. A. Alvi and A. Rahman, *Jahangir – The Naturalist*, New Delhi: National Institute of Sciences of India, 1968, pp. 4–14.

[1] Syud Ahmud, ed., *Tuzuk-i-Jahangiri* (henceforth, *Tuzuk*), Printed in Ghazipur and Aligarh, 1864.

birds and domesticated species. The wider scope of the *Memoirs* permitting only a limited space, he speaks of them but selectively. The rarer ones interest him more. Only when he has to say something new, or to give his own opinion with regard to a popular belief, does he touch upon the commoner species, leaving out the distinctive features as common knowledge.

Jahangir has his own manner of description. His interest in zoology emanated from a pervasive aesthetic bent of mind. Often enough his narratives are characterised by an overtone of admiration for the beauty he found in his subject. Nor does he hesitate to express his feeling of repulsion for an ugly-looking animal. Nevertheless, his descriptions are precise enough to make possible the correct scientific identification of the objects described. He gives local names, geographical distribution, even structural characteristics, weights, measurements and interesting notes on ecology and behaviour. His study of the sarus crane and the gestation period of the elephant represent two of his original contributions. In fact, modern zoologists, unaware of his work, had not correctly ascertained the latter phenomenon until the later half of the nineteenth century.

While the idea of any scientific classification was unknown to Jahangir, an understanding of groups is given out in as much as he uses the word *Alam* (literally meaning 'the world of', but implying a specific category) rather loosely for indicating comparable affinities of two or more animals which we classify under the same family and in good many instances, the same genus. Sometimes a group is indicated by analogy drawn of a particular feature (for instance, the bill, the size, the plumage, the coat of an animal or the like) with that of a closely allied species.

In botany, his interests were primarily horticultural. He tells us of having made it possible to cultivate high altitude trees like the cypress, juniper, pine and the Jawanese sandal tree in the plains of India.[2] He laid out some beautiful gardens. An interesting piece of information given by him pertains to the availability of mangoes and the preservation of other fruits until long after the close of seasons. He compares the fruits and grains of various regions and notes down average and record weights of some of the specimens. The aesthete in him is not dormant either.

[2] *Tuzuk*, p. 3.

The fragrant flowers of India catch his fancy at the very beginning of his narratives. He takes down detailed measurements of a date-tree 'which looked exceedingly-graceful', orders for a platform to be raised around it, and asks the painter to represent it in colour. Unfortunately, most of his paintings of plants, trees and flowers have been destroyed with the passage of time. The extent of this loss can be imagined from the fact that of the more than a hundred paintings of Kashmir flowers painted by Mansur alone, not one has survived to this day.[3]

His description of plants is not, however, very detailed. Generally he confines himself to specific aspects such as the trapping of the hornet in the lotus flower or the curious way the saffron sprouts from the soil.

The idea of groups is less evident in his treatment of plants. He identifies them by their fruits or flowers, or by their absence. Analogies given by him are superficial. He seems to be aware of some types of inflorescence but without realising their significance. The shapes and colours of the flowers interest him. In one case, he speaks of the stigma, in another case, of the stalk. But the relevance of the inner and outer components of the flower to its distinctive character is not understood. The shapes of the leaf and the stem casually draw his attention but not sufficiently to be of consequence.

A strong propensity for all that is wonderful, novel or phenomenal underlies his interest in other natural and scientific phenomena; and as he follows it up in the *Tuzuk* we are treated to a number of observations which could well have been worked upon towards really useful and conclusive information. There is a detailed description of plague, suggestive of the conditions under which it spreads. He speaks of a pair of elephants dying of rabies. Lunar and solar eclipses are recorded. The appearance and shape of a comet is meticulously observed over the period of a whole week.

Even in matters of day-to-day life, Jahangir displays a spirit of enquiry, which is more pronounced with respect to popular beliefs. He goes out of his way to measure the depth of a supposedly fathomless pond near Ajmer. Nowhere was it found to be more than 12 yards deep.[4] He despatches a team of his personal scribes with a surgeon to Banyan in order to check the authenticity of the belief that the body

[3] *Ibid.*, p. 300.

[4] *Tuzuk*, pp. 124–25.

of one Khwajah Yaqut, lying in a cave since the days of Mahmud Ghaznawi, had not decayed. The report is faithfully recorded. Half of the body that was in contact with the earth had undergone the natural process of decay. The rest was intact, including the nails and the hair of the head. The hair of the moustache and the beard on one side of the nose had fallen away. An epigraph, engraved on the entrance of the cave evidenced that the man had died before the days of Mahmud.[5]

There are numerous examples of this kind. These do not form part of the text of the present work. However, it would not be without interest to quote at least two of his experiments in his own words.

The following experiment was carried out to test the degree of the agreeability of the atmosphere of Ahmedabad and Mahmudabad, to human health. He writes:

> Undoubtedly the water and air of Mahmudabad have no resemblance to those of Ahmedabad. By way of testing this, I ordered them to hang up a sheep on the bank of Kankariya tank after taking off its skin, and at the same time one at Mahmudabad, that the difference of the air might be ascertained. It happened that after seven *gharis* had passed at that place (Ahmedabad) they hung up the sheep. When three *gharis* of day remained, it became so changed and putrid that it was difficult to pass near it. They hung up the sheep at Mahmudabad in the morning, and it was altogether unchanged until the evening, and began to be putrid when one and a half watches of night had passed. Briefly, in the neighbourhood of Ahmedabad, it became putrid in eight hours and in Mahmudabad, in fourteen hours.[6]

With regard to the healing power of *mumiya* (bitumen), he observes:

> I had heard much from physicians, but when I tried it, no result was apparent. I do not known whether physicians have exaggerated its effects, or whether its efficacy had been lessened by its being stale. At any rate, I gave it to a fowl with a broken leg to drink in larger quantity than they said, and in manner laid down by physicians, and rubbed some on the place where it was broken, and kept it there for three days, though it was sufficient to keep it from morning till

[5] *Ibid.*, p. 56.

[6] Henry Beveridge, ed., *Tuzuk-i-Jahangiri* or *Memoirs of Jahangir*, vol. II, trans. Alexander Rogers, London: Royal Asiatic Society, 1914, p. 33.

evening. But after I had examined it, no effect was produced, and the broken leg remained as it was.[7]

At the same time, Jahangir was often superstitious according to the fashion of time, thus his faith in astrology, then understood to be as real as any other science. But superstition went no further with him than taking auguries or carefully observing the advice of his trusted astrologers. It was co-existent and never confused with his innate rational sensibility. For instance, to him, the milk of a 'strange gelded goat' is as ominous as a shooting star but despite numerous witnesses, his intelligence would not agree to the widely believed story of a river having turned into a stream of milk.[8] Interested in mysticism, fond of the company of recluses, yet never intrigued by self-same demigods whose hoaxes he lays bare, he reveals a sagacious mind, resolved by its own disposition.

Jahangir's work may be appreciated better by keeping in mind a background of the state of science in contemporary Europe.

In the seventeenth century, science underwent a change of far-reaching consequences. It was then that the inadequacy of Aristotleian scholasticism, which had dominated scientific thought for 2,000 years, was substantiated and the first principles of experimental methodology were visualised. In this sense, this period has been rightly called 'the cradle of modern science'.

Essentially, however, it was a period of transition with all its characteristic merits and drawbacks. The system that we understand as science today was not perfected until after the eighteenth-century researches finally clarified the basic contents of the truly scientific methodology. While the achievements of such great men as Descartes, Vesalius, Harvey, Newton, Leibnitz and others definitely point to the new directions, in many respects, seventeenth-century ideas appear to be little better than the extension of the medieval concepts.

Natural sciences in general, and biology in particular, were the latest to respond to this change of outlook. Not until the publication of Linnaeus's *Philosophia Botanica* in 1751 and the 10th edition of his *Systema Natura* in 1758 were the principles of a scientific taxonomy laid down.

[7] *Ibid.*, vol. I, London, 1909, pp. 838–39.
[8] *Tuzuk*, p. 175.

While Linnaeus (1707–88) had some original thinkers among his immediate predecessors, biologists of the younger generation, more or less contemporary to Jahangir, did hardly better than perpetuate the knowledge and therefore, the ignorance of the past. Thus J. Johnston (1603–75), whose great zoological encyclopaedia was published in the 1660s, was merely copying Ulysses Aldrovandi (1552–1605) and Gesner (1516–65) and through them Aristotle in his classification of the vertebrates into viviparious quadrupeds, birds and fishes. Likewise his molluscs, testaceans, crustaceans, insects and zoophytes formed the category of 'bloodless' animals, the *anaima* of Aristotle. To him, as to his predecessors, bats were birds, the elephant was a soliped, the rhinoceros was a bisulc, the cetaceans were fishes, the amphibians and reptiles were oviparous quadrupeds, and the hippocampus (a mythical seamonster) was an insect.

Fable went hand in hand with fact and the literary history of animals passed for zoology. Superficiality of observation tended to degrade even the most painstakingly compiled works. Ill-conceived ideas about the habits of animals, their temperaments, cunningness, faithfulness, memory, usages, affection etc. occupy a greater space in their works than do zoological characteristics. The unicorn, the kraken, the basilisk, the harpies, the satyr and the stymhalid birds appear side by side with real animals in their descriptions. Aldrovandi thought that the barnacle goose sprang from the fruit of a tree. Thomas Moufet (d. 1604) believed that the king of bees (which in fact is a queen) was an elected head and could be dethroned by popular vote.

Fantastic ideas like these were not wanting in the study of Botany either. Andrea Cesalpino (wrote 1590–1603) likened a plant to an animal standing on its head. So that, it had a soul seated in the heart which was situated in its collar between the roots and the stem; the sap was carried to its various parts through arteries and veins; the roots were the plants, mouths and the flower grew in a foetus membrane much as the foetus of an animal. The classical doctrine of 'signature', according to which medical plants were supposed to bear organic resemblance with the diseases they cured, found great favour with the sixteenth century botanists and continued to be enthusiastically defended in the eighteenth century. Giambatista della Porta (d. 1595) even illustrated by a diagram the resemblance thought to be existing between the leaf of *Botrychium Lunaria* and the moon with the profile of a human face.

Classification was a matter of convenience with the medieval biologists, rather than a scientific principle. At that stage, they could not have possibly apprehended that animals and plants partake a kind of hierarchy in nature's scheme of evolution which is the true basis of taxonomy. They followed various systems: alphabetical, animistic (based on the biblical conception of an order in which God is believed to have created the animals on the fifth and sixth days of creation), subjective, such as based on culinary considerations, on the degree of domestication of animals or on the medicinal properties of herbs and plants. The comparatively rational systems such as were based on ecological, physical or generative analogies were uncritically followed in the tradition of Aristotle and his disciple Theophrastus. Instances of comparative studies in animal anatomy which, at any rate, were rare and unsystematic, are observed to have been carried out either to prove the 'Unity of God's plan' or as homage to the wisdom of Aristotle, rather than for taxonomical reasons.

But these biologists too had their own share of contributions to make. The great task they accomplished was to bring within the compass of biological knowledge, a great number of species of animals and plants.

Even as early as the beginnings of the Renaissance, botany and zoology had grown in popularity with scholars who rediscovered the ancient Greek and Latin texts and wrote commentaries on them. Geographical discoveries and the opening of new land and sea routes in the fifteenth century gave it a new impetus. A large number of animals and plants of the European countries were identified. Exotic species began pouring in from Africa, Asia Minor, India, America, the East and the West Indies. Some scholars like Pierre Belon (1517–64) and Andre Thevet (1503–92) performed hazardous voyages in order to study foreign fauna and flora in their natural habitats. By the turn of the sixteenth century, the catalogue of a few hundred species of animals and plants studied by Aristotle, Theophrastus, Dioscorides and Pliny had swollen up to the encyclopaedic volume of several thousands. The collected zoological works of Francisco Hernandez alone comprised 17 volumes. Among the great botanists, the names of Gherado Cibo who took eight years (1532–40) to complete his great herbal, Clusius (Charles de L'Ecluse, 1526–1609) who described 1,585 plants in his *Rariorum Plantarum Historia* (1601) and of the Bauhin brothers, John and Casper, are worth mentioning. John's encyclopaedia,

Historia Universalis Plantarum (1650–61) contained descriptions of 5,226 plants. Casper, the more famous of the two, described as many as 6,000 plants in his *Prodromus Theatrica Botanica* (1620). But Lobelius (Charles de l'Obel 1538–1616) did a unique job in his *Plantarum Seu Stirpum Historia* (last edition published by the author in 1581) which contained an index in seven languages. Linnaeus made extensive use of these works, returning to them time and again for reference.

Almost all the works produced in this period were illustrated, some exquisitely, by paintings, woodcuts or copperplate etchings, reproduced in print and published either separately in the form of albums or in the body of the main works. Belon was the first to publish his album. Clusius, Gesner, Silviani, Fuchs, Aldrovandi, the Bauhin brothers, all had their works profusely illustrated. John's encyclopaedia contained 3,426 figures. Casper illustrated all of his 6,000 plants.

To sum up, through the endeavours of the European scholars, biological knowledge grew rapidly in the sixteenth and seventeenth centuries. But the existence of myth and fable, so persistently observed in their works, suggests that the biologists of the period had not as yet developed even a rationale, if not a truly scientific attitude. There was one relieving factor that was to be of great historical importance: they worked with increasing vigour, continuing to build, brick by brick, with corporate endeavour, a consolidated tradition of up-to-date knowledge, which was the prerequisite to the ultimate discovery of the true system of biological thought.

To return to Jahangir, there should be no reason for one to feel cynical about the diminutive volume of his work, as compared to those of his Western contemporaries. Not all the enormous contents of the works of these scholars were entirely original, for they reproduced far more and often uncritically. Jahangir owed no debt to accepted authorities of whom there was no dearth in India. In the context of his times, however, he displays a good degree of rationality of approach.

As to its historical significance, Jahangir's work can by no means be said to represent the actual nature of contemporary scientific thought in India. A piece of unwitting scholarship, it was still the work of an amateur. But it does lead one to the speculation as to the worth of all that panorama of unrevealed history that lies buried on the shelves of the world libraries and museums. Unestimated numbers of specialised works on all the branches of science by Indian scholars of the past await study. Were they all written in vain, or is it possible to trace in them any direction towards the emergence of a true scientific philosophy? How did they lend themselves to the growth of a tradition, or was there no uniform growth at all? What is the nature and extent of their achievement or failure? What role did they play in the international scientific movement?

These are some of the questions Jahangir's work raises for us. A quest for the answers will, it is hoped, enlighten many a dark corner in the present picture of the world history of science.

Section II
FORESTS, PASTORALISM, NOMADISM

Forests, Pastoralists and Agrarian Society in Mughal India*

Chetan Singh**

The rural landscape of Mughal India is frequently depicted as a vast expanse of cultivated land peppered with innumerable villages. Its highly differentiated, peasant-centred village community has been graphically described by numerous scholars, along with the hierarchy of ubiquitous revenue officials upon whose diligence the Mughal political edifice ultimately rested. This picture of Mughal rural society, which has emerged through a laborious examination of court chronicles and official revenue records, is coloured by the prejudices of its original creators. What is implicitly articulated in this picture is a self-image of the very society that generated these chronicles and revenue records in the first place. Thus, subtly reasserted in many modern works, has appeared the world view of the Mughal courtier and the petty revenue official.

For the Mughals, a civilised society was one primarily engaged in agriculture, alongside other more sophisticated commercial activity. It was nurtured by, and conversely lent support to, a state that regarded the protection of the peasant as an essential obligation and the extension of agriculture as a cardinal objective. This perception was part of the socio-political thinking amongst dominant elements within the Mughal Empire. There was a sound underlying assumption here, namely that

* Chetan Singh, 'Forests, Pastoralists and Agrarian Society in Mughal India', in David Arnold and Ramachandra Guha, eds, *Nature, Culture, Imperialism: Essays on the Environmental History of South Asia*, New Delhi: Oxford University Press, 1995, pp. 21–48.

** I am grateful to the participants at the conference on 'South Asia's Changing Environment' held at Bellagio in March 1992. Their comments on an earlier version of this paper have helped me immensely. The long discussions I have had with Professor A. R. Khan, my colleague in the History Department at the University of Himachal Pradesh, have been equally useful.

the prosperity of the Mughal ruling class hinged upon its success in expanding and strengthening the empire's agrarian foundations.[1]

In reality, however, it was not entirely upon settled villages and revenue-yielding land that the stability of Mughal society depended. What seems to have escaped notice so far is the fact that there was much that lay beyond the agrarian economy of the Mughal heartland. If the Mughal 'system' in its process of expansion gradually created the social and economic context within which it functioned, it was in turn fundamentally affected by the 'others' whom it sought at least to dominate, if not to integrate. Over the long term, Mughal society and polity underwent several changes. In the making of these, no small part was likely to have been played by people and areas which were neither fully within the 'system' nor typically 'Mughal'.[2]

The Mughal ruling class did not enjoy an unchallenged monopoly over geographical space and natural resources even within the political confines of the empire. Ecological diversity ensured that these were shared with a host of other social organisations and utilised in different ways by each of them. It can be argued that environmental factors crucially affected the nature of socio-economic developments in the empire, and a proper appreciation of these factors would enable a better understanding of the evolution of state and society in large parts of pre-colonial South Asia.[3] This neglected issue is addressed here: what

[1] See Abu'l Fazl, *Ain-i-Akbari*, II, New Delhi: Munshiram Manoharlal Publishers, 1978, pp. 41–42, 46–50, 58–59, for the duties of the *faujdar* and *amalguzar*. Abu'l Fazl writes that the amalguzar 'should be a friend of the agriculturist…He should consider himself the representative of the lord paramount…He should strive to bring waste lands into cultivation and take heed that what is in cultivation fall not waste.' See also Irfan Habib, *Agrarian System of Mughal India, 1556–1707*, Bombay: Asia Publishing House, 1963, pp. 249, 251–56.

[2] For arguments regarding the role of tribes within the Mughal Empire, see C. A. Bayly, *Rulers, Townsmen and Bazaars: North Indian Society in the Age of British Expansion, 1770–1870*, Cambridge: Cambridge University Press, 1983, pp. 28–30, and his *Imperial Meridian: The British Empire and the World, 1780–1830*, Harlow: Longman, 1989, pp. 35–46; Chetan Singh, 'Conformity and Conflict: Tribes and the "Agrarian System" of Mughal India', *Indian Economic and Social History Review* (hereafter *IESHR*) 25: 3, 1988, pp. 319–40.

[3] Francois Bernier, *Travels in the Mogul Empire, 1656–1668*, New Delhi: Atlantic Publishers, 1989, p. 453. About the impact of environment Bernier

was the environment of Mughal India, and how did medieval society respond to it?

I

To begin with, it can be suggested that the extent of forested or uncultivated territory was considerable.[4] Accounts of Mughal forays against hostile local chieftains refer, all too frequently, to the thick jungles and inhospitable uncultivated countryside that the imperialists had to contend with during the course of their expeditions.[5] In the region of

---

observed, 'it is often said in the language of Aristotle that Egypt is the workmanship of the Nile, so may it be observed that Bengale is the production of the Ganges.'

[4] Shireen Moosvi, *The Economy of the Mughal Empire, c. 1595: A Statistical Study*, New Delhi: Oxford University Press, 1987, pp. 65–66; Shireen Moosvi, 'Ecology, Population Distribution and Settlement in Mughal India', *Man and Environment* 14: 1, 1989, pp. 109–16. During Akbar's reign the gross cultivated area was unlikely to have been more than 55 per cent of the gross cultivation in 1909–10 of large parts of north India. When compared to the geographical extent of the Mughal *subas* of this region, this would have been even less. Even Bernier's comment can be interpreted in this manner, for he informs us that 'of the vast tracts of country constituting the empire of Hindustan many are little more than sand or barren mountains, badly cultivated, and thinly peopled and even a good portion of the good land remains untilled for want of labourers ...' Bernier, *Travels*, p. 205. For a description of dense forests in the *terai* north of Gorakhpur, see Jean Baptiste Tavernier, *Travels in India*, II, New Delhi: Oriental Books Reprint Corporation, 1972 (first edition 1676), p. 205. See also Habib, *Agrarian System*, pp. 1–24; Irfan Habib, *An Atlas of the Mughal Empire*, New Delhi: Oxford University Press, 1982.

[5] There are also several earlier accounts which show quite clearly that large forested areas were a common feature in north India. During his campaign in the Siwaliks against hill rajas, Timur had to cut his way through thick forests: he even ordered that while preparing for battle the soldiers should carry hatchets to 'clear away the jungle': *Mulfuzat-i-Timuri or Tuzuk-i-Timuri* (the autobiography of Timur) in H. M. Elliot and J. Dowson, *The History of India as Told by its own Historians*, III, rpt. New Delhi: Kitab Mahal, 1964, pp. 463, 469. See Ziauddin Barani, *Tarikh-i-Firoz Shahi* in Elliot and Dowson, *History of India*, III, pp. 103–06, for a description of the thick jungles encountered

Chanderi, for instance, Babur's artillery was preceded by 'active overseers and a mass of spadesmen to level the road and cut the jungles down…'[6] During the time of Akbar, the campaign against Raja Madhukar found the Mughals advancing towards Orchha in a similar fashion. We are told this about the surrounding territories: 'the country was forest, and the marching of the army was difficult, they cut down the trees one day and marched the next.'[7] Even during subsequent reigns, the military campaigns in this region were conducted in this manner.[8]

In many parts of Bihar, too, a similar problem was encountered by the Mughal armies. Shahbaz Khan, who was assigned the task of suppressing the rebellion of Gajpati, a *zamindar* of Jagdishpur in Bihar, had to spend 'nearly two months engaged in cutting down the trees' around the fort before he could capture it.[9] It could be argued that these forts were, for strategic reasons, deliberately built in thickly wooded and isolated areas and that information pertaining to military campaigns does not necessarily indicate the forested nature of large tracts of territory. While this might be partly true, there were nevertheless many fortresses which were located in, or near, well-populated towns of reasonable importance where the approaches were well forested. A case in point was Kangra. We are told by DeLaet, an early seventeenth-century traveller, that the only approach to Kangra was 'through a forest 50 *cos* broad', and that

---

during campaigns by Balban. Regarding a campaign in Katehar, it is noted that 'Woodcutters were sent out to cut roads through the jungle, and the army passing along these brought the Hindus to submission': Abdul Qadir Badaoni, *Muntakhab-ut-Tawarikh*, rpt. New Delhi: Idarah-i Adabiyat-i Delli, 1973, p. 378.

[6] Zahiruddin Muhammad Babur, *Baburnama*, rpt. New Delhi: Oriental Books Reprint Corporation, 1970, p. 572.

[7] Abu'l Fazl, *Akbarnama*, rpt. New Delhi: Rare Books, 1972, III, pp. 324–25.

[8] Abdul Hamid Lahori, *Badshahnama*, in Elliot and Dowson, *History of India*, VII, p. 48, for the campaign against Jajhar Singh Bundela at Chauragarh (near Jabalpur) in 1634.

[9] *Akbarnama*, III, p. 261; *Muntakhab-ut-Tawarik*, II, p. 182. For the area around Gaya, see *Akbarnama*, III, pp. 472–73. See also Nuruddin Muhammad Jahangir, *Tuzuk-i-Jahangiri*, rpt. New Delhi: Munshiram Manoharlal Publishers, 1968, I, pp. 315–16, for the campaign against Durjan Sal of 'Khokhara' in Bihar.

while advancing on the fort the Mughals had to cut down the forest as they advanced at a rate of 'about a *cos* a day', and 'finally reached the fortress in the eighth month'.[10] The fortresses of Jasrota and Mau were similarly located in 'impenetrable jungles'.[11] This seems to indicate that many *qasbas* (rural market towns), towns and some adjoining agricultural tracts were still in the nature of clearings in the midst of undisturbed forests. Interestingly, in a letter to the Shah of Iran, Akbar recounted amongst his major achievements his advance into Kashmir, during which thousands of men were employed in 'removing rocks, and in cutting down forests and making roads'.[12]

While their writ ran unchallenged over large agricultural expanses, the Mughals found it difficult to cope in places where forest, desert or topography sheltered strong political entities from their pervasive and expanding influence. Contemporary observers repeatedly refer to areas that were not easily amenable to Mughal control.[13] Manucci, who

[10] Joannes DeLaet, *The Empire of the Great Mogol*, Bombay: D. B. Taraporevala Sons and Co., 1928, p. 195. A much earlier reference to the town (*shahr*) of Nagarkot which is a large and important town of Hindustan says that 'the road thither lay through jungles': *Malfuzat-i-Timuri*, in Elliot and Dowson, *History of India*, III, p. 465.

[11] Shaikh Illahadad Faizi Sirhindi, *Akbarnama*, in Elliot and Dowson, *History of India*, VI, pp. 127–28.

[12] *Akbarnama*, III, p. 1010. See also DeLaet, *Empire*, pp. 183–84, regarding a Mughal campaign in Rajasthan during Jahangir's reign: 'the royal army then advanced to Siavend ... a very strong place, which the Kings of Delhi had never dared to attack on account of the impenetrable wilderness and forests by which it is surrounded. Abdul Chan gave orders that these forests should be gradually cut down ahead of the advancing army...' On reaching the fortress, 'The moat was filled up with logs of trees, and so huge a mound was raised against the walls that at last, the garrison... fell.'

[13] Apart from the more powerful rulers who ruled over different parts of India, Babur noted that 'there were also, in the hills and jungles, many *rais* and rajas': *Baburnama*, p. 481. Mahomed Kasim Ferishta, *History of the Rise of Mahomedan Power in India till the Year* 1612 CE, rpt. New Delhi: Munshiram Manoharlal Publishers, 1981, IV, p. 313 emphasises Babur's words: 'India abounds with forests and extensive wilderness, full of all sorts of trees; so much so, that these wastes seem to offer inducement, both to rajas and subjects, to revolt from the government' DeLaet, *Empire*, p. 242, says that 'the empire contains many provinces which are rendered difficult of access on account

lived and travelled in India for a long time, makes a very interesting observation. He writes:

> There are also in this empire other lords who call themselves zamindars ... Such men do not maintain cavalry: the greater number live in the midst of jungles, and these usually pay no revenue, unless it be taken by force of arms. At this day, taking the whole Mughal empire, these rajas great and petty and the zamindars, exceed five thousand in number.[14]

While periodic military expeditions ostensibly brought many such areas under imperial paramountcy, the Mughals were still inclined to allow them a large degree of economic and political autonomy.[15] This relative freedom enjoyed by the chieftains was not merely the outcome of a proclaimed policy of magnanimity. The Mughal state, in fact, did not have the military and administrative wherewithal to directly govern and simultaneously obtain economic benefit from these areas of apparently low productivity.[16] John Fryer observed during the reign

---

of their mountainous character and the dense forests with which they are covered. Large armies cannot operate in such districts, which are held by Radias [rajas]... If opposed by the Mogols with a greater force than they can cope with, they merely retreat into their mountains and await a better opportunity of success.' Cf. Berner, *Travels*, p. 205: 'The empire of the Great Mogol comprehends several nations, over which he is not absolute master.'

[14] Niccolao Manucci, *Storia Do Mogor*, rpt. Calcutta: Editions Indian, 1965, II, p. 417. Cf. *ibid.*, II, p. 414: 'The roads are not direct, owing to the forests and mountains and the interposition of territories belonging to the different rajas and zamindars who allow no travellers to pass through, out of the fear they have of the Mogul.'

[15] For a detailed description of the political relationship between the chiefs and the Mughal Empire under Akbar, see A. R. Khan, *Chieftains in the Mughal Empire during the Reign of Akbar*, Simla: Indian Institute of Advanced Study, 1977.

[16] Though the Mughal army was successful against Durjan Sal in Bihar, Jahangir says that 'in consequence of the difficult roads and thickness of the jungles they contented themselves with taking two or three diamonds and left him in his former condition': *Tuzuk-i-Jahangiri*, I, p. 315. DeLaet, in *Empire*, p. 243, sees the Mughals' inability to control these areas as a serious political problem: 'The Mogol Emperor has hitherto proved unable to find a cure for these dangerous diseases of the body politic.'

of Aurangzeb that the Mughal armies were 'unfit for such barren and uneasy Places …'[17] Moreover, the monetary worth of such territory could not easily be converted into the agricultural revenue statistics to which Mughal revenue officials were more accustomed.

Therefore, the continued existence of autonomous chieftaincies within the Mughal Empire was to a great extent the result of the geography and physical environment. If on the one hand the vulnerability of larger cultivated stretches to Mughal military might have facilitated close supervision by revenue officials, it was the existence of 'jungles' and 'ravines' within the territory of many of the chieftains, on the other, that probably discouraged direct Mughal interference. While the finances of several of these chiefships drew heavily upon peasant-centred agricultural activity,[18] the proportionally large uncultivated areas within their boundaries must also have necessitated the search for economic alternatives. A wider mix of productive activities utilising the resources of forests and grassland may thus have become essential. The relatively greater freedom that peasants probably enjoyed in such an environment may sometimes have been incentive enough for many of them to migrate from the more rigidly administered imperial territories to those controlled by autonomous rajas.[19] Each chieftaincy, influenced as it was by its own peculiar environmental situation, must have developed a suitable balance of a variety of economic activities from which it derived its revenues. Examples of this sort were the principalities located in Rajasthan and Gujarat. According to Ferishta, in the territories of these chiefs, 'little other grain but *bajry* and *jowar*

[17] John Fryer, *A New Account of East India and Persia being Nine Years' Travels, 1672–1681*, reprinted Liechten: Kraus Reprint Limited, 1967 (first edition 1698), II, p. 59. This comment was made with reference to the Mughal campaign in the Deccan against Shivaji.

[18] Most of the larger and powerful chiefships maintained a formal state organisation in which an important part was played by the land-revenue administration. Two relevant regional studies are G. D. Sharma, *Rajput Polity: A Study of Politics and Administration of the State of Marwar, 1638–1749*, New Delhi: Manohar, 1977, and Dilbagh Singh, *The State, Landlord and Peasants: Rajasthan in the Eighteenth Century*, New Delhi: Manohar, 1990.

[19] Francisco Pelsaert, *Remonstrantie*, published as *Jahangir's India*, New Delhi: Idarah-i Adabiyat-i Delli, 1972, p. 47; Bernier, *Travels*, p. 226; Habib, *Agrarian System*, pp. 336–37.

are cultivated. The revenue is for the most part derived from horses and camels'.[20]

Any South Asian state attempting to expand to subcontinental dimensions would, therefore, have been confronted with a rich diversity of both physical and socio-economic environments. Furthermore, if it endeavoured to go beyond merely establishing an overarching political suzerainty and sought to squeeze out surplus directly from local-level economies, such a state would have had to evolve a multiplicity of revenue-extracting methods, each suited to a particular socio-economic system. While it had the ability to penetrate down to lower social levels in many agricultural areas, the Mughal revenue machinery relied considerably on intermediaries for its efficient functioning. In regions that were strikingly dissimilar from those where the Mughal revenue officials operated, the autonomy of such intermediaries must have been even greater. This meant that the Mughals were able to obtain a share of the wealth generated in divergent and peripheral economies only in an indirect manner. Very often, a substantial part of the *peshkash* (tribute) that many of the autonomous chiefs sent to the emperor consisted either of forest produce or pastoral products which implies the harnessing of grazing land.[21] Despite their inability to exercise direct

[20] Ferishta, *Mahomedan Power*, IV, p. 318; *Ain-i-Akbari*, II, p. 256.

[21] An earlier reference to how this non-monetary system of tribute extraction worked is obtained from an account of Kabul by Babur for 1519. He fixed the tribute of the Khirichli and Samu-khail Afghans at '4,000 sheep'. A punitive raid on the Waziri Afghans compelled them to proffer a tribute of '300 sheep': *Baburnama*, p. 413. Cf. Zain Khan, *Tabaqat-i-Baburi*, rpt. New Delhi: Idarah-i Adabiyat-i Delli, 1982, p. 110; *Ain-i-Akbari*, II, p. 183 for honey, wax and wooden-ware from the mountains north of Awadh; DeLaet, *Empire*, p. 59, refers to valuable herbs from the territory of Raja Basu (Nurpur in Kangra). *Tuzuk-i-Jahangiri*, I, p. 218: the peshkash of the Raja of Kumaon included *gunth* ponies, hawks, falcons, 'navels of musk', swords (*khanda*), daggers (*katar*) etc. According to Bernier, *Travels*, pp. 419–20, some areas in Kashmir paid the annual tribute in leather and wool; Tavernier, in *Travels*, II, p. 206, says the Raja of Nepal sent an elephant in tribute annually. See Habib, *Atlas of the Mughal Empire*, p. 41, for elephants taken in tribute from the Morung chiefs. Manucci, *Storia*, II, pp. 415–16, says—with regard to the smaller rajas who had strongholds located north of the Ganges—that they sent the 'rarities produced in their country. Some send gavioens (sparrow-hawks),

economic control over these areas of mixed (agrarian, pastoral, forest) economies, the Mughals were thus able to extract from them, over irregular periods, a small portion of their wealth in the form of peshkash. This was perhaps the best available compromise that the Mughal state could strike, through the chieftaincies, with diverse socio-economic systems that were largely different from its own.

Like the Mughal Empire, however, most of the autonomous chiefships which interacted with it were reasonably developed state formations. This enabled the Mughals to establish an acceptable political relationship, the principles of which were in fact formulated as early as the reign of Akbar.[22] Amongst these were also to be found tribal chieftaincies that had accepted Mughal suzerainty or even been incorporated into its military structure. Yet in many parts of Mughal India, there existed several tribes which could not be made to conform consistently even to a loosely-defined political relationship.[23] It was these more fiercely autarchic tribes which probably lay even further from the Mughal 'system'—as it has come to be understood by scholars. Many such tribes were, from time to time, to be found in conflict with Mughal authority and had established a reputation as 'thieves and plunderers' of the 'king's territories'.[24] The topography of the areas they occupied suggests that they relied considerably, though not exclusively, on pastoralism. With this they combined a sense of social or communal

---

falcons and other birds of prey, pretty birds, honey and wax. With the last article they produce a waxed cloth for the lining of tents and other uses in the royal household. They also make candles for the harem...'

[22] Khan, *Chieftains in the Mughal Empire*, pp. 206–24.

[23] Bayly, *Rulers, Townsmen and Bazaars*, p. 80: 'in this period the state was only one of the political formations...The land-revenue based state and settled agriculture did not occupy the whole of the social and geographical space of the area, and there were large tracts where they were both on the defensive.'

[24] Manucci, *Storia*, II, pp. 428, 430. Cf. Singh, 'Conformity and Conflict' pp. 337–38. Forest areas in medieval England were often viewed as being disorderly, as breeding 'atheism and consequently all disobedience of God and King': John Martin, *Feudalism to Capitalism: Peasant and Landlord in English Agrarian Development*, London: Macmillan, 1983, p. 200. See Bayly, *Rulers, Townsmen and Bazaars*, p. 219, for British attitudes to tribes in colonial India, where 'Nomads and wanderers were seen as disorderly elements—carriers of roguery and dissidence.'

cohesion that was characteristically distinct from the differentiated agricultural village communities of the core Mughal areas.

Scattered references in our sources indicate that societies which were predominantly non-agricultural in nature were viewed as primitive and as a threat to the settled agrarian areas where Mughal land revenue regulations were methodically applied. The strangeness of some of these to Mughal society comes through to us from a few stray comments of Abu'l Fazl. Near the fort of Chunar (in *suba* [province] Allahabad), we are told, 'is a tribe of men who go naked living in the wilds, and subsist by their bows and arrows and the game they kill.'[25] While hunter-gatherers certainly appeared peculiar to contemporary observers, there were many other tribes which, even though partially agricultural and well known to Mughal society, were also perceived as being socially distinct. The *koli*s, *bhil*s and *gond*s (suba Khandesh) are referred to as people who 'can tame lions so that they will obey their commands' and about whom 'strange tales are told'.[26]

The gonds, in particular, are repeatedly mentioned in the *Akbarnama* on account of their considerable population and the large area they controlled.[27] Apart from discovering that the gonds lived 'in the wilds', we also learn that 'the people of India despise them and regard them as outside the pale of their realm and religion'.[28] Quite clearly, therefore, these peripheral societies were considered by the Mughals as external to their own society. Yet, because they were seen as an occasional hindrance to Mughal hegemony, the imperial authorities intermittently endeavoured to establish a military dominance over these people as a prelude to more fundamental economic inroads into their territories.[29]

[25] *Ain-i-Akbari*, II, p. 170.

[26] *Ibid.*, II, p. 233. See *Muntakhab-ut-Tawarikh*, II, p. 178, for the *Kharwaha* tribe of sailors, and W. H. Moreland, *India at the Death of Akbar: An Economic Study*, rpt. New Delhi: Reprints and Trans Publication, 1974, p. 27.

[27] *Akbarnama*, II, pp. 323–24. According to 'faithful narrators', Garh Katanga 'contained 70,000 inhabited villages'.

[28] *Ibid.* Their low status did not, however, prevent Salbahan of Ratha, a Chandel raja with declining fortunes, from giving his daughter Durgavati in marriage to Dalpat, the son of Aman Das, a wealthy gond chief.

[29] *Ibid.*, pp. 563–64. In 1582 Todar Mal proposed regulations by which Mughal functionaries could take effective action against the 'dwellers in the ravines' who, being of a 'turbulent disposition think the ruggedness of their

Also significant is the fact that such societies were not confined to the geographically distant parts of the empire. Some of them were located quite close to important political and military centres,[30] and probably combined the practice of agriculture with a considerable dependence upon pastoralism.[31] What seems to have insulated them from Mughal authority, irrespective of distance, were the densely forested areas which usually surrounded, or were interspersed with, their territories.[32] It was into these forests that they retreated when confronted with stronger Mughal armies. In the arid areas where animal husbandry was a primary occupation, the paucity of water and the inadequacy of local food supplies severely restricted the movement of large invading armies and, therefore, discouraged permanent Mughal occupation. Their own economies being structurally suited to the desert, the pastoralists

---

country a protection and make long the arm of oppression'. The aim was to reduce them to a docile revenue-paying peasantry, and nomads were to be settled as agriculturists. In the context of Kabul, Abu'l Fazl notes: 'There are many wild tribes… and most of them at the present time have become settled colonists': *Ain-i-Akbari*, II, p. 407. Akbar proudly refers to the suppression of the 'carnivorous Afghans' and the 'wicked Baluchis and other desert-dwellers who are of a bestial nature…': *Akbarnama*, III, p. 1010.

30 Proper control over the fort of Rohtas in Bihar could only be established after the 'bandits of the neighbourhood of the fort' had been coerced by the Mughal army: *Akbarnama*, II, p. 287; cf. *ibid.*, p. 252, where Abu'l Fazl says that 30 *kos* from Agra, near the town of Sakit, were to be found lawless elements who 'were both ruffians and occupiers of rough places'. The Jats and *Mewatis* who were to be found at the very centre of the Mughal Empire are well-known examples.

31 Cf. the observation regarding the Roman legions which in the time of Caesar encountered Germanic tribes that were 'settled agriculturists, with a predominantly pastoral economy': Perry Anderson, *Passages from Antiquity to Feudalism*, London: Verso, 1978, p. 107.

32 *Ain-i-Akbari*, II, pp. 251–52, 254, for the koli and *jaitwah* tribes living in the forests; *Tuzuk-i-Jahangiri*, II, p. 285, for 'villagers and cultivators', who 'passing their time in the shelter of thick jungles and difficult strong places in stubbornness and fearlessness, would not pay their rents to the *jagirdars*'; DeLaet, *Empire*, pp. 19, 21, 34, 184, for bhils, 'Grassias' and kolis, who lived in 'solitary places' and 'trackless retreats'; Manucci, *Storia*, II, p. 428, for Jats; see also Sujan Rai Bhandari, *Khulasat-ut-Tawarikh*, New Delhi: Rare Books, 1918, p. 63.

could seek temporary refuge in terrains where the Mughal forces dared not venture.[33]

At this point, however, it is important to sound a note of caution against overemphasising the autonomy of the pastoral tribes. They were not totally isolated socio-economic entities, and their distinctiveness from agrarian society did not prevent them from interacting with it. In the north-west, many Afghan tribes, particularly the *lohanis*, were activity engaged in trading activities that drew them into a fairly close relationship with settled and commercialised society.[34] Some sections among the kolis, bhils, gonds, *bhattis* and probably numerous other tribes were significant elements of the Mughal military force in certain areas.[35] Despite their socially peripheral position, the various tribes of herdsmen scattered all over the Mughal Empire would have engaged in various kinds of economic exchange with the adjoining agrarian society. Pastoral products, particularly ghee, were items of commercial value and served as an important linking factor between the two economies.[36]

[33] DeLaet, *Empire*, pp. 68–70, referring to the area around Thattah, mentions 'savages who recognize no ruler' and were engaged in robbing travellers. Furthermore, 'if the king sends his armed forces against them they burn their huts, which are made of straw, and withdraw to rugged mountains.' See also Khan, *Chieftains*, pp. 63–73.

[34] *Baburnama*, p. 235; *Akbarnama*, III, p. 1160; DeLaet, *Empire*, pp. 69–70. See B. R. Grover, 'An Integrated Pattern of Life in the Rural Society of North India During the Seventeenth and Eighteenth Centuries', *Proceedings of the Indian Historical Records Commission*, 1966, p. 5, for *gujars*.

[35] *Ain-i-Akbari*, p. 233. While the kolis, bhils and gonds constituted the provincial military force of suba Khandesh, they do not seem to have been part of the peasantry in this otherwise well-cultivated suba. Manucci, *Storia*, II, p. 430, for bhattis of Lakhi Jangal in suba Lahore entering the faujdar's service.

[36] See *Ain-i-Akbari*, I, p. 60, for large quantities of ghee from Hisar-Firoza used in Akbar's kitchen. *Raibari* herdsmen were also apparently employed to graze the royal camels; *Ain-i-Akbari*, I, pp. 155–56. Shiv Das Lakhnawi, *Shahnama Munnawar Kalam*, Patna: Janki Prakashan, 1980, relates an incident where a caravan was plundered by Jats near Hodal during the time of Churaman Jat: the merchants were 'carrying with them 1300 carts of clarified butter ' which were taken away by the robbers. Ghee in such huge quantities was very likely to have been procured from pastoralists. See also Chetan Singh, *Region and Empire: Panjab in the Seventeenth Century*, New Delhi: Oxford University Press, 1991, pp. 266–70; Bayly, *Rulers, Townsmen and Bazaars*, p. 29.

It was probably from such tribes that new social groups were added to the sedentarised agricultural population as and when the need arose.[37]

While he was resident in Kabul and prior to the invasion of Hindustan, Babur seems to have depended to a fair extent upon the pastoralists of that area; even subsequently, when he was near the Salt Range, he ensured that the flocks of country people were not harmed.[38] During their early years in India, the Mughals were probably better able to utilise the economies of pastoral societies. Apart from the fact that the extent of Babur's kingdom was much smaller than that of his descendants, the Mughal land revenue system was yet to emerge in its fully institutionalised form. Quite possibly, then, the smaller size and the relative lack of organisational rigidity of the state both required and facilitated the tapping of non-agrarian sources of income from scattered pastoral communities. After its establishment on firmer 'agrarian' foundations and subsequent expansion, however, the Mughal Empire seems to have relied primarily upon land revenue from large agricultural expanses. With this shift in the scale and nature of the state's financial activity, extraction of economic surplus from scattered pastoral societies probably became a distracting task with unpredictable results. Despite these fundamental changes, however, the interaction between Mughal agrarian society and the pastoralists can hardly have ceased.

As suggested earlier, the relationship between the two was not simply one of outright confrontation. Apart from the apparent political and military dimensions of the relationship, its socio-economic content seems likely to have been far more substantial than has so far been recognised. In many parts of the Mughal Empire, the lives of peasants and pastoralists were inextricably entwined. Even while being involved in the mutual exchange of products, the two were simultaneously engaged in a silent and fluctuating struggle of encroachment upon and retreat from each other's living space. While agriculturists frequently intruded into areas where the herds of pastoralists grazed, it was equally common for abandoned fields and habitations to be reappropriated by herdsmen who wandered on the fringes of cultivated areas.[39]

[37] Singh, *Region and Empire*, pp. 269–70.

[38] *Baburnama*, pp. 380, 402.

[39] Delaet, *Empire,* pp. 46–47, says, of Sikandarabad, near Bayana, that the place is in ruins and 'is inhabited only by a few shepherds called Goagers

The history of agriculture in Mughal India was not, therefore, a case of the unabated expansion of cultivation and the relentless destruction of forests. There were times and places where the forest reasserted itself; old clearings were abandoned and new ones made in different places. This is apparent from an observation during the reign of Akbar by Fath Ullah Shirazi regarding revenue matters: 'The fluctuations of civilization are apparent to everyone. If in a village some land falls out of cultivation one endeavours to increase cultivation elsewhere.'[40]

II

> It ought to be remembered that the whole of the merchandise which is exported from the Moghul kingdom, comes from four kinds of plants—that is to say, the shrub that produces the cotton from which a large quantity of cloth, coarse and fine, is made…The second is the plant which produces indigo. The third is the one from which comes opium, of which a large amount is used on the Java coast. The fourth is the mulberry tree, on which their silk-worms are fed, and, as it may be said, that commodity [silk] is grown on these trees.
>
> – Niccolao Manucci[41]

By stating matters in this way, Manucci reduced the most commercialised sector of the Mughal economy to an essential dependence upon 'four kinds of plants'. In so doing, he implicitly provided a partial explanation for the pronounced interest of the Mughals in fostering agriculture and championing the cause of the peasantry against more mobile social groups.

Since the primary objective of the state was, as we have seen, the extension of cultivation, the peasants were encouraged to break new land. Upon the inhospitable fringes of agricultural areas were probably to be found ploughmen who laboured to push back the forest. Certain land revenue regulations were specifically formulated as an incentive for this purpose. That newly cultivated land was entitled to concessional

---

(Gujars)'. Even the 'vast ruins of the ancient city of Delhi' had become 'only an abode of shepherds'.

[40] *Akbarnama,* III, p. 690.

[41] Manucci, *Storia,* II, p. 393.

rates of taxation is well known. In addition to the state, influential members of the village community played an important part in the granting of ownership rights on virgin land to those who had the capital resources to bring it under cultivation.[42] The village *muqaddams* (village headmen) were given the authority to parcel out uncultivated village land to whoever offered to bring it under the plough.[43] In the case of *madad-i-maash* grants (revenue-free land grants given to men of learning and religious persons), too, half the land granted consisted of cultivatable waste which the grantee was expected to bring under cultivation.[44] Given the incentives that were provided by the Mughals, it is very likely that many *zamindaris* emerged at the frontiers of cultivation as a result of the clearing of forests organised by entrepreneurs of this kind. The actual work of felling was done by groups of professional woodcutters and there probably existed, in some regions, entire castes which specialised in this kind of work.[45]

Bankatai (the clearing of forests) and the subsequent settling of cultivators were adequate reasons for staking a claim to the zamindari of an area, a claim which imperial authorities readily recognised.[46] Adjacent

[42] Satish Chandra, 'Writings on the Social History of Medieval India: Trends and Prospects', p. 89, and 'Role of the Local Community, the Zamindars and the State in Providing Capital Inputs for the Improvement and Expansion of Cultivation', pp. 166–83, both in *Medieval India: Society, the Jagirdari Crisis and the Village*, New Delhi: Macmillan, 1982.

[43] Habib, *Agrarian System*, p. 133; Ali Muhammad Khan, *Mirat-i-Ahmadi*, Baroda: Oriental Institute, 1965, p. 241. In the *farman* dealing in great detail with the matter of land revenue, it is stated that where 'land is not owned, the owner is unknown, it should be given to one who is able to reclaim it so that he reclaims it.' See Bayly, *Rulers, Townsmen and Bazaars*, p. 101, for certain castes of cultivators which were engaged in the extension of agriculture in the eighteenth century.

[44] *Ain-i-Akbari*, I, p. 280; Habib, *Agrarian System*, pp. 302–03.

[45] Chandra, 'Role of the Local Community', p. 175 John Francis Gemelli Careri, 'A Voyage Round the World', in J. P. Guha, ed., *India in the Seventeenth Century*, New Delhi: Associated Publishing House, 1979, II, p. 311: 'There are also two tribes of Sutars or timber-men; the one called Concanes, and the other Gujarati.'

[46] Muzaffar Alam, 'Eastern India in the early Eighteenth Century Crisis: Some Evidence from Bihar', *IESHR*, 28: 1, 1991, pp. 65–66. In suba Bihar, ploughs were donated by the government to such people. 'Every *hal mir* (i.e.

to the core agricultural areas of the Mughal Empire, therefore, was territory where a perpetual struggle went on between cultivators and forests. These were also areas where different socio-economic systems interacted with each other. They were driven to this as much by the logic of an expanding agrarian economy as by the pastoralists' or forest-dwellers' need for essential commodities from settled agriculturists.

Even at the best of times, and in close proximity to many of the agriculturally developed territories, there existed only an uneasy balance between cultivated areas, cultivable village wasteland and forests. Whenever the struggle between agriculturists and the jungle intensified, or conversely when the husbandmen were forced to abandon cultivation, even this equilibrium was overthrown. It is quite likely that the oppressive demands of revenue officials occasionally compelled peasants to leave their homes and seek shelter in the forests.[47]

An ambivalence, therefore, seems to have marked the peasant's interaction with the jungles that were often adjacent to his fields. His hostility towards the untamed wilderness was tempered by a basic dependence upon its natural abundance. Timber for the construction of his hut[48] and implements obviously came from these forests, as did fuel and much of the cattle fodder.[49] Now and then, when wild animals

---

one who has four or five ploughs) was to be 'given a *dastar* [turban] so that he may clear forests and bring land under cultivation'.

[47] During the reign of Muhammad Tughlak, when taxation in the Doab became unbearable, the cultivators 'set fire to their houses, and retired to the woods with their families and cattle. Many populous towns were abandoned and remained so for several years.' Furthermore, certain tribes which 'inhabited the country about Soonam and Samana, unable to discharge their rents, fled to the woods,' Ferishta, *Mahomedan Power,* I, pp. 243–44.

[48] *Baburnama*, p. 428, about the nature of villages in Hindustan: 'they need not build houses or set up walls—*khas*-grass abounds, wood is unlimited, huts are made and straightway there is a village or town!'

[49] B. Ch. Chabra, ed., *Antiquities of Chamba*, New Delhi: Manager of Publications, 1957, II, p. 147. The Chamba plate of Balabhadra (dated 1641) refers to '*banj* trees used for fodder'. *Mirat-i-Ahmadi,* p. 256, refers to 'leaves of *dhakah, paliyah* and *babul* skin etc., brought from the jungle …' Though the reference is to their sale in the city, it seems reasonable to assume that the villages, with easier access, would have utilised such forest resources quite freely. Moreland, *India at the Death of Akbar,* p. 144, argues that while the rural

were successfully hunted down, a little extra food was obtained.[50] Interestingly, even salt was obtained, in Assam, by processing the large leaves of a 'fig-tree'.[51]

It was in times of dire need, however, that the importance of forests came to be fully realised. An early eighteenth-century description of famines in the regions of Palwal and Hodal brings out the crucial role of forests as a source of additional food. During the famine of 1713, we are told, 'Many repaired to jungles and collecting the leaves and blossoms of the *Karial* boiled them and subsisted on the same.'[52] In the following year, the rains came rather too late to benefit cultivation, but were sufficient to rejuvenate the vegetation in the forests. We learn that 'In the forest a kind of wild millet (*shamakh*) shot up so profusely that the people of that region, big and small, went daily to the forest and collected as much as each of them could bring, and thus gave themselves some energy.'[53] Thus, as a last resort, the forest was often of critical importance to famine-stricken people, a resource that stood between life and death.[54]

Recognition of the value of forests and their usefulness to the rural economy is implied in the land grants made by Raja Balbhadra of Chamba in 1641. To his cook the raja gifted the village of Kuhmaro

---

population enjoyed an 'unrestricted supply of such produce', this advantage was offset by 'the damage caused to their crops by the wild animals'.

[50] *Ain-i-Akbari,* II, pp. 181–82, for wild buffaloes which were numerous in suba Awadh. Abu'l Fazl says that 'people find sport in hunting them'. Wild asses were found in some of the drier areas and were, according to Jahangir, 'lawful food': see *Akbarnama*, II, p. 522; *Tuzuk-i-Jahangiri*, I, pp. 83–84.

[51] Tavernier, *Travels*, pp. 221–22.

[52] Lakhnawi, *Shahnama Munnawar Kalam*, p. 22.

[53] *Ibid.*

[54] Armies that ran short of food supplies during military campaigns were forced to search for food in forests. See *Muntakhab-ut-Tawarikh,* II, p. 242. In a campaign against Mewar the Mughal army had to 'sustain life upon the flesh of animals and the mango fruit. This latter grew there in such abundance as defies description. The common soldiers used to make a meal of it... in default in bread.' Cf. *Akbarnama*, III, p. 682: when Muzaffar Gujarati fled Mughal forces in 1548 he went to the 'mountains of Barda' where the forests 'thirty *kos* long and ten broad, are well-watered, and produce abundance of wild fruit'.

inclusive of the 'forest and the hill-slope...'[55] In another grant of the same year the village of Sarotha was gifted to a brahmin along with a 'watermill as well as the *banj* [oak] trees used for fodder'.[56] The fact that forest land and trees were considered worthy of specific mention in land grants suggests that in certain areas various kinds of rights over them had crystallised. Some of these rights, indeed, appear to have taken the form of individual ownership.

Pastures were another very important natural resource for the peasantry. But because of variations in different areas and at different seasons, it would be inappropriate to make a definite statement about the amount of pastureland that was available. A stray remark by Abu'l Fazl to the effect that the means of earning a livelihood were 'as abundant to the labourer as forage for his cattle' conveys the impression that there was no shortage of pasturage.[57] This impression is further reinforced by the fact that during four months of the rainy season, and also when the emperor was on the march, no fodder allowance was permitted for animals in the imperial stables.[58] Bernier, on the other hand, in a more detailed account, seems to contradict Abu'l Fazl by referring to a 'great deficiency of pasture land'. 'The heat', he noted, 'is so intense and the ground so parched, during eight months of the year, that the beasts of the field, ready to die of hunger, feed on every kind of filth like so many swine'[59]. In his description of the countryside, perhaps around Surat, another European traveller similarly observed that as early as the month of March, while the leaves on the trees were green, the grass was 'quite burnt up'. He goes on to add that though the plain areas were rich in all kinds of natural produce, the dry summers caused a shortage of pasturage.[60] For a large part of the year, therefore, the leaves procured from the neighbouring forests probably constituted a substantial portion of the cattle fodder.

With regard to pastures, however, it seems quite likely that the rights of different social groups were better defined than was the case with

[55] *Antiquities of Chamba*, II, p. 146.
[56] *Ibid.*, p. 149.
[57] *Ain-i-Akbari*, II, p. 56.
[58] *Ibid.*, I, pp. 152, 160.
[59] Bernier, *Travels*, p. 326.
[60] Fryer, *New Account,* II, pp. 94–95.

the forests. Control over pastureland was central to the economy of the tribes which were predominantly pastoral. In suba Kabul, for instance, where agriculture was not very productive,[61] the tribal economy apparently depended on pastoralism. Two of the most powerful tribes here were the Hazaras and the Afghans, and Abu'l Fazl makes a point of mentioning that 'the pasturage of the country is in the hands of these two clans'.[62] The grazing lands of the Hazaras were virtually treeless meadows, supporting grass particularly suitable for horses and for the large number of sheep they kept.[63] Nearer the town of Kabul it appears that the Mughals demarcated separate grazing areas for the animals of different sections of their army. A dispute arising from an infringement of these could escalate into a major confrontation.[64]

In well-cultivated villages, the extent of pastureland was likely to have been relatively little. This is what seems to be implied by the statement in the *Ain-i-Akbari* that tax would have to be paid on cattle by 'whosoever does not cultivate land liable to taxation but encloses it for pasturage'.[65] The need to convert cultivated land to pasture indicates that grazing land used in common by the villages was not always adequate. The peasants were therefore particular about where the village livestock was permitted to browse. Grazing rights were perhaps fairly well developed, and the ordinary cultivator was not entirely free to let his cattle roam all over the village commons.[66] Indeed, in certain areas,

[61] *Ain-i-Akbari*, II, p. 405.

[62] *Ibid.*, II, p. 406, The Hazaras, in particular, were probably more dependent upon horses, sheep and goats.

[63] *Babarnama,* pp. 222–23. Not all grassland in Afghanistan was suitable for rearing livestock. Babur points out areas where the grass, despite growing in abundance, was good neither for horses nor sheep.

[64] Mutamad Khan, *Iqbalnama-i-Jahangiri*, in Elliot and Dowson, *History of India,* VII, p. 428. 'A party of Rajputs turned out their horses to graze in the hunting ground near Kabul and a contention arose with the keepers, in which an *ahadi* was killed ... the *ahadis* sought redress... and dissatisfied with the answer they received...attacked the Rajputs, and killed 600 or 700 of them.'

[65] *Ain-i-Akbari*, II, p. 49.

[66] DeLaet, *Empire*, p. 95: 'The condition of the peasants is such that they have to approach the chief man of their village (who is appointed by the king) and to declare to him what land, and how much they intend to sow, or where they will pasture their flocks and herds.'

individual usufructuary, if not proprietary, rights over clearly-defined categories of uncultivated land had already emerged. The Himalayan state of Chamba was one such area, and numerous land grants dating from as early as the tenth century until Mughal times indicate this.

Many of these inscriptions specifically mention the inclusion of 'grass, grazing and pasture-ground' in the grant.[67] Furthermore, the officials within whose jurisdiction the land lay were ordered to desist from cutting the grantee's 'pasture whether green or ripe' or from seizing 'his wood, fuel, grass, chaff and so on'[68]. Apart from land that was thus donated by the raja, there appear to have existed larger stretches of fairly well-delineated pastureland. There were, in fact, some reserved pastures for which the term *ghali* was used and the extensive territory of Brahmaura (the homeland of the Gaddi shepherds) was referred to as a ghali.[69]

It was perhaps on the wastelands located at some distance from the carefully cultivated areas that large numbers of cattle were allowed easier access.[70] But with regard to such places, too, it might be erroneous to suggest that grazing rights were completely unrestricted. The existence of reserved pastures even in regions of natural abundance, such as Chamba, shows that this was not the case. It suggests, furthermore, that the dangers of pasturelands becoming depleted were real even in medieval times.[71]

[67] J. Ph. Vogel, ed., *Antiquities of Chamba State*, Calcutta: Superintendent of Government Printing, 1911, I, pp. 168, 189, 196, 200; II, p. 69. See also Sharma, *Rajput Polity*, p. 86. Taxes on grazing (*ghasmari*), along with other agricultural cesses, were assigned to the grantee.

[68] Vogel, *Antiquities,* I, p. 169.

[69] *Ibid.*, II, p. 126, Nagoda rant of Balbhadra, 1634 CE. See also p. 183, for a glossary defining the word ghali. During the course of describing the boundaries of the land-grants, there are frequent references to such land bordering on the 'pasture-land' of particular areas; e.g., p. 38, the 'pasture-land of Mugala.'

[70] Fryer, *New Account,* I, p. 287, of the Surat region says, 'Here in the Marshes are brought up great store of cattle of all sorts...'

[71] For a comparison with European areas similar in many ways to Chamba, see Georges Duby, *Rural Economy and Country Life in the Medieval West*, London: George Allen and Unwin, 1969, pp. 146–47: the 'opening of the "alps", and the meadows on the high mountains was organized' during the thirteenth

A question that arises in this context is whether the common cultivator in the Mughal period had a larger stock of cattle for carrying out his work than did the peasant of a later time. On account of the relatively greater extent of uncultivated land and forests, pastures and fodder were, it seems, more easily available to the Mughal peasant. The *Ain-i-Akbari* also states that while collecting tax on cattle, an exemption of four oxen, two cows and one buffalo per plough should be allowed to the cultivator.[72] On the basis of this exemption, it has been argued by some scholars that the peasant in the Mughal period maintained more cattle than his twentieth-century successor.[73] It may, however, be erroneous to arrive at such a conclusion. Among the primary objectives of the state was, as noted earlier, the extension of cultivation. The liberal tax exemption that the Mughals allowed on cattle is to be seen as an incentive in this direction and not as an indicator of the number of cattle that an ordinary peasant owned. For this reason, the tax exemption was linked 'to each plough'.

Studies pertaining to eastern Rajasthan provide some interesting details in this regard.[74] While some of the prosperous landholders certainly owned numerous heads of cattle, the large majority, lower down the social scale, were ordinary peasants who possessed only one or two oxen.[75] Caste and other factors within the village community may have prevented the poorer sections from becoming landowners, and they

---

century, and 'on the alpine slopes which had until then been uninhabited, the lords would install at great altitudes herds of 50 to 100 cattle... and entrust them to families of herdsmen.' By 1345 the high pastures on the Alps had been overgrazed and the village communities in the hills attempted to prevent the entry of outsiders.

72 *Ain-i-Akbari*, II, p. 49.

73 Moreland, *India at the Death of Akbar,* p. 106; Habib, *Agrarian System*, pp. 53–55; Moosvi, 'Ecology', p. 114.

74 Chandra, 'Role of the Local Community', and 'Writings on Social History', in *Medieval India.*

75 In *qasba* Chatsu (1606 CE), for instance, 79 per cent of the cultivators had one or two oxen; in *mauza* Kotkhawda 68.7 per cent. Of 26 villages in *pargana* Chatsu, 68.7 per cent of the peasants had one or two oxen. The percentages for peasants in this category in the 94 villages of Chala Kalan (*sarkar* Rewari) and the 49 villages of Kotla (sarkar Tijara) were 79.65 and 58.74 respectively. In qasba Chala Kalan as a whole it was 45.8 per cent. See Chandra, 'Role of the Local Community', pp. 178–79, 182–83.

would thus have maintained very few cattle of their own. Moreover, the wasteland available for supporting livestock in the immediate vicinity of well-cultivated villages in the agrarian areas was not inexhaustible. From available information it seems, therefore, that only a few rich peasants had four or more oxen and that the cattle population of the medieval agricultural village was not really as large as it has so far thought to have been.[76] Under the circumstances, one is tempted to argue that if the number of cattle per capita during the Mughal period was, indeed, greater than what it was in the early twentieth century,[77] then it was not merely on account of the large number of livestock possessed by the peasantry. A very substantial part of it was owned by pastoralists whose huge herds grazed in the innumerable pastures or found abundant fodder amidst thickly-forested tracts.

III

The existence in many areas of an abundance of forests and pastureland should not convey the impression that Mughal society was unfamiliar with the scarcity of forest produce and fodder. Such an impression would be misleading, for in many cultivated regions, natural resources were already under stress. In some areas, for instance, fuel had to be transported over long distances. During the reign of Akbar, 600 wagons were set aside for bringing 'in the space of ten months, 1,50,000 *mans* of fuel to the Imperial kitchen'.[78] Even in Kashmir, one of the obligations of the cultivator was to 'cut and bring some wood from a distance'.[79] The conveyance of firewood on boats in Kashmir, to

[76] *Ibid.*, pp. 178, 182–83. In qasba Chatsu only 10 per cent of the peasants had four or more oxen, and in the 26 villages of pargana Chatsu an average 16.78 per cent of the peasants had four or more oxen.

[77] In *Agrarian System*, pp. 53–54, Habib argues that there was a 'larger number of working cattle per head of population'.

[78] *Ain-i-Akbari*, I, p. 159. One Akbari *man* was equal to 55.32 lbs. Babur makes what seems a contradictory statement about Kabul. He notes that 'excellent firewood is had nearby' and then adds: 'Given one day to fetch it': *Baburnama*, p. 233.

[79] *Akbarnama*, III, p. 1088. For references to the transportation of firewood to cities, see *Mirat-i-Ahmadi*, pp. 148, 232, 256.

which Jahangir refers, seems to indicate that the distances involved were considerable and that wood for fuel was not easily available in the vicinity of habitations.[80] Even in parts of Punjab 'firewood and grass' seem to have been similarly transported.[81] The scarcity of fuelwood is explicitly mentioned by Pelsaert when he writes that 'firewood is very dear and is sold by weight', and that the 'poor burn cow-dung mixed with straw dried in the sun…'[82] The latter, too, was sold!

Nor was the procuring of fodder for animals always a simple task. A large number of horses, camels and elephants were maintained by the Mughal emperor and the *mansabdars* (nobles) who resided in the cities. At the imperial stables, cash allowances were made for the purchase of hay, grass and oats, and it is also likely that land was set aside to provide fodder for these animals. The price of fodder was naturally higher in the cities compared to the countryside.[83] That a reasonable profit could be obtained by people dealing in this commodity is implied by references in our sources to professional grass-sellers.[84] Some of them were required to ensure its regular supply to the imperial army wherever it proceeded, and Bernier has provided an interesting description of how they went about it:

[80] *Tuzuk-i-Jahangiri,* II, p. 142.

[81] *Ibid.*, I, p. 66. During his rebellion, Khusrau attempted to cross the Chenab at Sodharah. However, he could only lay his hands on two boats, one of which was 'full of firewood and grass'.

[82] Pelsaert, *Remonstrantie,* p. 48. The use of 'dried cowdung' is further substantiated by DeLaet, *Empire,* p. 89, and by Bernier, *Travels,* p. 368. See Fryer, *New Account,* I, p. 111, for a toll levied on cow-dung by the English in Madras.

[83] *Ain-i-Akbari,* I, pp. 142–43, 152. We are informed that 'three *bighas* of land will yield sufficient fodder for a horse', and that 'in winter each horse gets a *bigha* of fresh oats'. Iraqi and Turki horses at the court were provided a daily allowance of two *dams*, and in the countryside one and a half dams. The price of oats also varied.

[84] *Tuzuk-i-Jahangiri,* I, p. 432. We come across a reference to a 'grass-seller' in Ahmadabad killed by a thief who had broken into his house. Lakhnawi, *Shahnama Munnawar Kalam,* p. 8: when ordered to capture some 'free-booters' who had been creating trouble near Araul (40 miles southwest of Patna) a Mughal official mistakenly arrested 'thirteen men, some of whom were vegetable and grass vendors'.

> These poor people are at great pains to procure forage: they roam about from village to village and what they succeed in purchasing, they endeavour to sell in the army at an advanced price. It is common practice with them to clear, with a sort of trowel, whole fields of a peculiar kind of grass, which having beaten and washed, they dispose of in the camp at a price sometimes very high and sometimes inadequately low.[85]

There were times when the shortage of fodder would become acute. This was particularly the case in the eighteenth century, when large armies moved through well-populated areas. We have the interesting example of the invading army of Nadir Shah which, after making numerous assaults on the Mughal encampment near Karnal, carried away 'corn, grass and wood'![86] During Ahmad Shah Abdali's confrontation with the Marathas at Kunjpura, the shortage of fodder became so severe that from time to time, 'one or two thousand of the Maratha horse went out for grass and forage'.[87] Abdali was subsequently able to prevent such excursions; matters reached such a state that the only people who ventured out of the Maratha camp were 'wretched naked labourers who, going by stealth into the open country, used to dig up grass from the ground with their *kharpas* and offer it for sale'.[88]

It seems that, in as early as the latter part of the sixteenth century, the cost of construction materials in urban areas was rising. Perhaps on this account, Akbar was prompted to fix the prices of different materials.[89] Among these was wood. Abu'l Fazl provides a detailed price-list of eight kinds of wood, graded according to quality and size.[90] Some of the other

[85] Bernier, *Travels*, p. 381. Shops selling 'forage' were a regular part of the tent-bazaar which sprang up wherever the imperial army encamped.

[86] Rustam Ali, *Tarikh-i-Hindi*, in Elliot and Dowson, *History of India*, VIII, p. 61.

[87] Muhammad Jafar Shamlu, *Tarikh-i-Manazil ul Futuh*, in Elliot and Dowson, *History of India*, VIII, p. 150.

[88] *Ibid.*, p. 151.

[89] *Ain-i-Akbari,* I, p. 232: 'Many people are desirous of building houses; but honesty and conscientiousness are rare, especially among traders. His majesty has carefully inquired into their profits and losses, and fixed the prices of articles…'

[90] *Ibid.*, p. 233.

materials included in this price regulation were bamboo, *khas*, reeds for thatching, *munj* and *san*.[91] Traders must have brought these materials into the larger towns and cities in huge quantities. From Bernier's description of Delhi, we learn that even amongst the high Mughal officials, very few possessed houses 'built entirely of brick and stone'.[92] The city, in fact, consisted of an immense number of thatched cottages which gave Delhi the appearance of a 'collection of many villages'.[93] In one year, fire is said to have destroyed 60,000 thatched-roof cottages in Delhi. Considering that the materials mentioned above had already become items of commercial value, even constructing houses made of bamboo or wood was not always cheap.[94]

By the seventeenth century, at least, timber of good quality seems to have become a commodity which could not be obtained very easily. In Punjab, where it was used for boat-building as well as house construction, it had to be floated along rivers which issued out of the northern hills.[95] While clearing the thick jungles around Jasrota (near Jammu) during a campaign, the Mughal commander Shaikh Farid is said to have come across 'several old trees' which, being suitable for purposes of construction, 'were cut down and sent to Lahore for use in government buildings'.[96] The growing commercial value of timber led to its organised exploitation in forests that were easily accessible. Some of the Himalayan chieftaincies rich in this resource, too, seem to have benefited financially on this account.[97]

[91] *Ibid.*, p. 234. *Kah-i-chappar* (reeds for thatching) were sold in bundles. Munj was 'the bark of *galam* reeds and was used for making ropers to fasten the thatching'. San was mixed with quick-lime and ropes were also made of it.

[92] Bernier, *Travels,* p. 246.

[93] *Ibid.*

[94] *Ain-i-Akbari*, II, p. 134. In Bengal a single bamboo house could cost Rs 5,000: *Muntakhab-ut-Tawarikh,* II, p. 185. In Bihar a type of house called '*chappar-band*' could cost as much as '30,000 or 40,000 rupees each, although they are covered with wood'.

[95] Bhandari, *Khulasat-ut-Tawarikh*, p. 77; Habib, *Atlas of the Mughal Empire,* p. 32: Khizrabad, on the banks of the Yamuna, was a timber mart.

[96] Sirhindi, *Akbarnama*, in Elliot and Dowson, *History of India,* VI, p. 126. According to Babur, *mahua* wood was used most in the houses of Hindustan: *Baburnama*, p. 505.

[97] *Sirmur State Gazetteer, Part A, 1904,* Lahore: Punjab Government Press, 1907, p. 14. A farman (dated 1084 H) of Aurangzeb is quoted to the effect

There had emerged, therefore, in certain parts of Mughal India, a profitable market for forest produce and other similar natural resources.[98] Some of these were transported over fairly long distances, suggesting thereby that they were not available everywhere, or not easily available in adequate quantities. In many of the typically 'Mughal' agrarian areas, growing commercialisation and the extension of cultivation combined with the growth of urban centres to put a strain on locally accessible materials.

But there still remained the peripheral regions within the territorial bounds of the Mughal Empire where such resources existed in plenty. It was into these areas that agrarian society steadily encroached, both pressurised and supported as it was by the Mughal state. The combined logic of military expansion, increased cultivation and the growing requirement of well-populated areas for natural resources[99] brought the Mughal state into even closer contact with societies residing in the forested and relatively less cultivated areas.

Agrarian society was certainly not the sole beneficiary of this increased interaction. Many peripheral societies which had earlier been only marginally affected by the Mughal state were gradually drawn into playing an increasingly important role in its economy as well as its

---

that the emperor asked Raja Budh Prakash of Sirmur to 'permit a contractor to take *sal* timber from the Kalakhar forest free of charge and to refund to him any due which had been levied. Timber worth Rs. 8000 was, in consequence, taken for imperial use.' It is obvious, therefore, that forests had become a source of revenue for some chiefs and that contractors were engaged in the business of timber extraction.

[98] For the existence in medieval times of a market in Himalayan forest produce, see Chetan Singh, 'Humans and Forests: The Himalaya and the Terai during the Medieval Period', in A. S. Rawat, ed., *History of Forestry in India*, New Delhi: Indus Publishing Company, 1991, pp. 171–72.

[99] Apart from the collection of wood, herbs, grass etc., cows and elephants were also captured from the forests: *Travels of Thevenot*, in Guha, *India in the Seventeenth Century*, II, pp. 66–67. 'The huntsmen of Agra go five days' journey from the town as far as a mountain called Narwar… their business in going so far is only to catch a kind of wild cow which they call *merous*, that are found in the wood round this hill, which is upon the road from Surat to Golconda; and these cows being commonly very lovely, they make great advantage of them.' For cultivators paying revenue in elephants, see *Ain-i-Akbari*, II, p. 207.

political fortunes. These societies were neither numerically nor militarily insignificant, and their active participation, either as fuller constituents of the empire or as its opponents, could hardly have been without consequence. It seems reasonable to suggest, all things considered, that the ecological diversity found within Mughal India enabled and encouraged the existence of a corresponding variety of socio-economic systems. Among these, the most influential was an agrarian society based principally upon peasant production. Growing interaction between diverse social systems affected not only the peripheral areas and their people, but also transformed the nature of agrarian society and the Mughal Empire which was the latter's most powerful political expression in South Asia.

The Baiga*

Survival Strategies and Local Economy in the Central Provinces

Archana Prasad**

This essay makes an attempt to revise dominant notions about the relationship of forest communities and their livelihood systems to the local and regional agrarian economies. The term 'forest communities' is used for groups of people who depended primarily on forests for their daily requirements in the Central Provinces during the nineteenth century. For the most part, environmental historians have assumed that these communities have lived in isolation from, and have conflicting interests with, the local economy. In this essay, I challenge this assumption and show how the livelihood systems of one community, the *baiga* (conventionally regarded as a group of swidden cultivators) of Eastern Central Provinces, was dependent on the complex interconnections between the local and regional economies. The parameters of the local economy were in turn defined by relations of production and exchange that constituted these networks. By overlooking these relationships in their analysis, scholars have often tended to overemphasise the role of practices like shifting cultivation in the study of the identity and material life of these communities. In the process, they have also tended to romanticise the relationship between people and forests.

* Archana Prasad, 'The Baiga: Survival Strategies and Local Economy in the Central Provinces' in *Studies in History*, volume 14, issue number 2, July–December 1998, pp. 325–348.

** **Acknowledgements:** Parts of this essay were presented at the University of California, Berkeley, and Department of Science and Technology, Cornell University in 1995–96, and at the Nehru Memorial Museum and Library in 1997. I am grateful to those who commented on this essay in these places. In particular, I thank Neeladri Bhattacharya, William Storey, S. Sivaramakrishnan and Mahesh Rangarajan for detailed comments.

A simplistic connection between culture and ecology, often referred to as 'cultural ecology', is evident in the twentieth-century descriptions of baiga life and livelihood of which Verrier Elwin's *The Baiga* is a good example. For example, Elwin, as we shall see in this essay, goes out of his way to prove that cultural practices correspond to ecological roles. In the process, he overstresses the importance of the eco-friendly character of *bewar* (the baigani term for swidden) which was practised in the thick *sal* forests and the laterite soil of the Maikal hills. Scholars like Ramachandra Guha and Mahesh Rangarajan also assume that baiga swiddening was a 'sustainable' form of resource use.[1] They stress that the logic of bewar was firmly embedded in the consciousness and mythology of the community. Therefore, ecological roles and practices of the community were regulated by community codes. Conversely, cultural practices also reflected the ways in which the baiga reacted to the ecological compulsions that delimited their survival options. The resultant 'cultural ecology' is based more on anthropological particularism than on any historically valid notion of the community and change.

Arguments such as these need to be re-evaluated if historical evidence is to lead to more realistic and alternative forest and 'tribal' policies. In this context, I attempt to discuss the usefulness of Elwin's *The Baiga*, as a source in the reconstruction of baiga lifestyles in the late nineteenth and twentieth centuries. Second, I seek to reconsider the relationship between culture and ecology in baiga life. This is done in the context of the wide-ranging changes in the agrarian structures of the local and regional economies. In the process, I aim to widen the scope of the questions that have hitherto been posed by a number of 'environmental historians'. That changes in the composition of forests had a profound effect on the livelihood strategies of the baigas is evident, but whether this factor was primarily responsible for the marginalisation of the baigas is debatable. It has been my contention that the state monopoly over *sal* forests was a manifestation of larger issues relating to the integration of the baigas into the wider political economy, and this integration was based on a structural relationship

[1] Mahesh Rangarajan, *Fencing the Forest: Conservation and Ecological Change in India's Central Provinces, 1860–1914*, New Delhi: Oxford University Press, 1996.

between forestry, agriculture and rural industry within a given local economy.[2] Thus, no study of the process of marginalisation can be content with looking at either forestry or the environment as a self-contained field of study.

Forestry and the Problem of Swidden Cultivation in Baiga Life

The baiga of the Central Provinces depended on a combination of gathering, hunting and swidden cultivation practices for their survival in the early nineteenth century. Of these, swidden cultivation was considered the cornerstone of baiga livelihood. To this end, the debate over the ecological and economic viability of bewar, a baigani term for shifting cultivation, dominated discussions of baiga survival. By the first quarter of the twentieth century, anthropologists like Elwin used the argument that bewar was both eco-friendly as well as essential to the preservation of baiga values. Since bewar was being practised since times immemorial, it represented the psychological security and material basis of baiga life.[3] As Elwin saw it, 'tribal values' can be more humane, and therefore morally superior, to the exploitative values of modern 'civilised society'.[4] By making such an argument, Elwin was attempting to confront and disprove the statements of imperial officials who believed that bewar was an expression of the material and cultural backwardness of baiga life.

[2] For a detailed analysis of this in the Central Indian context see Archana Prasad, 'Forests and Subsistence in Colonial India: A Study of the Central Provinces, 1830–1945', Unpublished Ph.D. dissertation, Centre for Historical Studies, Jawaharlal Nehru University, 1994.

[3] I have re-evaluated the proposition that the forest communities were dependent on forests and swiddening from times immemorial in my 'Military Conflict and Forests in Central Provinces, India: Gonds and the Gondwana Region in Pre-colonial History', *Environment and History* 5: 3, October 1999, pp. 361–75.

[4] For Elwin's defence of tribal values see Verrier Elwin, *Aboriginals*, London: Oxford University Press, 1940. For Elwin's defence of bewar and its role in the baiga psyche see his introduction in *The Baiga*, New Delhi: Oxford University Press, 1939.

Elwin's argument has had wide historiographic impact. Rangarajan picks up these arguments in favour of others, and presents them as self-evident historical truths. He writes:

> Bewar was always practised on forest lands and also on hill slopes. On the other hand *dhya* could be practised outside the woods and was often taken up in the plains. The gonds cut their *dhyas* in the forests but often commenced ploughing in the second year.[5]

Like Elwin, Rangarajan attempts to simplify the confusion of the records by asserting the cultural and regional distinctiveness of the baiga's bewar. Thus, the study of bewar becomes an exercise in the expression of cultural identity rather than a study of the community's survival patterns. In this way, the complex nature of the ecological viability issue is neglected by historians of modern India. For example, Guha accepts the analysis of Elwin when he writes: 'shifting or jhum cultivation was a characteristic form of cultivation over large parts of India, especially in hilly and forested tracts where plough cultivation was not always feasible'.[6] While Guha and Gadgil are quick to question the prevalence of plough agriculture in forests, shifting cultivation is not subject to the same scrutiny. Neither do they question the proposition that emphasises the role of bewar as the only important survival activity in baiga life.

This section explores the relationship between shifting cultivation, hunting and gathering practices of the baigas. The survival of bewar cultivation depended intrinsically on the links between the baigas and the rest of the local agrarian economy. There are very few nineteenth-century sources for the reconstruction of the structural parameters of such seasonal activities. Part of the problem lies in the nature of these colonial sources, which often project the strategies as 'destructive to the forests'. Unpublished proceedings of departments, settlement reports and other records betray a broad bias towards sedentary permanent cultivation. However, this alone is not the limitation of these nineteenth-century sources. Their distinguishing feature is their confusion over the classification of these practices.

[5] Rangarajan, *Fencing the Forest*, p. 102.

[6] Madhav Gadgil and Ramachandra Guha, *This Fissured Land: An Ecological History of India*, New Delhi: Oxford University Press, 1992, p. 150.

For example, most nineteenth-century official records use the term bewar for both the field as well as the system of cultivation. But they also use the term *dhya* for this purpose. Reports of officials like Richard Temple often use these two terms interchangeably. There is also a third angle to this classification: the word dhya is also used for shifting cultivation in general without making a distinction between the baiga form, bewar, and the Gond form, *burra*. The difference between gond cultivation forms and the practices of the baigas is only mentioned in the reports of some foresters. The attempt to wish away the problem of agricultural land classification in forested areas reflected the confused images of baiga swiddening in nineteenth-century official records. The centrality of bewar in the late nineteenth- and early twentieth-century baiga society reveals that the mobility of the baigas had become a crucial impediment to the achievement of colonial aims in the forested areas.

At the time of the formation of the Central Provinces in 1861, the survival system of the baigas consisted of a loosely structured system of three different strategies: gathering, hunting and bewar. These three were bound together in a precarious balance by a fragile seasonal rhythm, whose success was intrinsically dependent on the success of the bewar cycle. Unlike other forms of permanent cultivation, bewar was only cultivated on the same field for three consecutive years, the cropping patterns and agricultural cycle differing from one year to the next. In the periods when no cultivation took place, gathering and hunting became the basis of baiga survival. For example, in the first year the baigas prepared the bewar field and, therefore, their dependence on hunting and gathering was likely to be greater than in the second year, when the fruits of the first crop were available. Gathering and hunting were seasonal and undertaken when bewar was not possible. Consequently, the periods of hunting and gathering were determined by the bewar agricultural cycle shown in Table 4.1.

Table 4.1 shows that there were two periods of gathering and hunting given the prevalent bewar cycle. Women gathered *mahua*, roots, *sal* seeds and fuelwood, while men were busy firing, preparing and working on the fields. The men indulge in different kinds of gathering activities when the lac cocoons need to be picked or when the *ral*, tan or dye tapping season arrives. Scholarship on the baigas and other forest communities has not taken these hunting and gathering activities as an intrinsic part of seasonal survival. Instead, they have considered them as auxiliary or supplementary activities.[7] Yet, if we analyse the linkages between these three activities it is possible to surmise that the shifting cultivation system is difficult to sustain without the existence of hunting and gathering.

Table 4.1: *The Agricultural Cycle under Bewar*

Indian Calendar	*English Calendar*	*Nature of Work*
Chait	March–April	Clearing fields of undergrowth and large trees
Baisakh	April–May	Firing of the *bewar* field
Jeth	May–June	Broadcasting of seeds
Asad	June–July	Men work in fields
Sawan	July–August	Men work in fields
Bhadon	August–September	Men work in fields
Kuar	September–October	Beans ripen and are harvested
Kartik	October–November	*Kodon* (a species of paspalum) and *kutki* (a small perennial herbaceous plant) ripen
Aghan	November–December	All crops are harvested
Pus	December–January	Winnowing takes place
Magh	January–February	Shift to new *bewar* fields
Phagun	February–March	Mostly festival time

Source: Drawn by Author.

The example of the *khonds* of Orissa, a forest community whose survival strategies are similar to the baigas, may be used to make this connection more explicit. In Chait, women went to clearings to pick castor and cut stalks that were already ripened. Khond women gathered fruits like sago, tamarind and mushrooms. Baiga women gathered only

[7] For this point see Rangarajan, *Fencing the Forest*; James Forsyth, *The Highlands of Central India: Notes on their Forests and Wild Tribes, Natural History and Sport*, London: Chapman and Hall, 1889; Robert Vane Russell and Hira Lal, *Tribes and Castes of the Central Provinces of India*, vol. 1–4, London: Macmillan & Co., 1916.

roots and mahua (*Bassia latifolia*) seed at this time.[8] While women did the gathering, men cut large trees and prepared the bewar fields. They also went on their ritual hunt which began at full moon. The baiga were known as the best hunters amongst all forest communities in the Central Provinces. They used bamboo bows, arrows and spears. The spears were made of iron and sometimes consisted of a bamboo shaft with an iron head. The bows and arrows were generally made of *dhamin* (*Grewia elastica*) wood. Traps were only used to protect the fields.[9] In Baisakh, the firing of the forest took place, and the women gathered unburned wood to burn. Men continued to hunt, but only in nearby forests. The seeds were sown in Jeth, and hunting continued. From Asadh to Bhadon, men worked in the fields. In Kuar, the first beans were ripened and in Kartik, small millets or *kutki* became ripe. In Aghan every crop was ready and in Pus winnowing took place. Pus was also the time for dances and marriages. In Magh, shifts were made to new bewars and hunting–gathering was the main source of survival. This was the agricultural cycle of the first year. In the second year, there was more time for hunting because fewer crops had to be sown and harvested, but since there was enough food, the people lived on the bewars. It was only in the third year that the diet had to be supplemented with the forest products and game from hunting.[10]

The Khond case shows the dynamic nature of seasonal survival in non-sedentary cultivation practices. It indicates that the dependence of the community on forms of survival other than cultivation was maximised in the first year. In that year, bewar had to be prepared for the first time. The cultivators were occupied in clearing the fields and dependence on hunting and gathering activities was almost total. The interaction between bewar and other seasonal activities was minimised in the second year. This was primarily because the community could depend on the crop from the first year. During this period, hunting

[8] Verrier Elwin Papers, Number 165, Notes on the Khonds, Nehru Memorial Museum and Library, p. 89 (hereafter *Khond Notes*). Here, Elwin uses some valuable nineteenth-century monographs to make his case. He makes a similar case using his anthropological experience in *The Baiga*. Since the connections in that monograph are ambiguous, I use the Khond experience to show the dynamic nature of the swiddening cycles.

[9] Elwin, *The Baiga*, p. 87.

[10] *Khond Notes*, p.109.

and gathering were used for both ritual purposes and for supplementing the diet. But crop from the first year remained the basis of survival. In the event of its failure, the community depended solely on hunting–gathering. In the third year, a combination of both cultivation and hunting–gathering was the basis of survival. The communities depended upon the crops of the second year and fruits, roots, etc. from the forest. Thus, the three-year cycle of survival was intrinsically dependent on the success of the crop of the first year. If it failed, the precarious balance between these three modes of production would be disturbed.

It is therefore evident that the importance of bewar in the baiga livelihood system varied from year to year and from one season to the next. This was true even as bewar formed the fulcrum of the seasonal cycle and enjoyed a dialectical relationship with hunting and gathering. Without them, bewar itself would become unsustainable. Consider, then, some of Elwin's contentions regarding the nature of bewar:

> I have quoted enough to show [from historical evidence] that there is among those who have served the Baiga the best …, that given proper conditions the forest always recovers itself in time, that given a small tribe like the Baiga the effect on erosion and rainfall was not a very serious danger.[11]

Elwin's plea is both moral and ecological in character. His argument for the restoration of the baigas' right to bewar is ecological. However, the claim that bewar was eco-friendly and did not hinder the production of timber trees is not entirely correct. It involves an inquiry into several questions. What does Elwin mean by the term 'forests'? Does his perception of the forest match that of the foresters? Do the baigas see the forest in the same way as Elwin? And finally, is the representation of bewar as a self-sufficient system correct? The answers accrue from analysis of the relationship of bewar fallows with the regeneration of timber trees like sal and the role of these trees in baiga survival.

The north-eastern Maikals, the principal locus of bewar cultivation, had an abundance of thick mixed forests. Early travellers to the area described it as a 'wild' environment characterised by undulating plains and laterite soil. A combination of the nature of the slopes, soils and forests

[11] Elwin, *The Baiga*, p.129.

provided the setting for bewar, gathering and hunting. The sustenance of these activities was intrinsically dependent on a nine-year fallow period which each field was allowed after one year of cultivation. The main aim of these fallow cycles was the regeneration of forest undergrowth in order to replenish the nutrients of the soil. If the undergrowth failed to appear, it was impossible for the cultivator to do bewar cultivation on the field. What swiddening did not require was the growth of large timber trees. In fact, there is sufficient evidence to show that the cultivation cycle often interfered with the full regeneration of sal trees that had been harvested in the preparation of the bewar. The frequent complaint of colonial foresters that bewar resulted in the replacement of mature trees with young trees of a stunted growth was sometimes justified.

Early surveys of the region by foresters show that good sal was found on gentle undulating slopes of Mandla district. Foresters described the maximal conditions for the regeneration of the sal tree as rainfall between 40 and 80 inches and good sub-soil drainage. This was possible on a surface with permeable rocks. The penetration of the surface soil was ensured by slopes in places where the soil was not very good. But where the soil drainage was good, the best sal would grow also on low lying plains.[12] The rainfall in the Maikals was about 56 inches, grasses were lush and helped to retain some moisture in the ground hardened by the thick canopy of the sal. Trees like bamboo and sal had a high rate of natural regeneration. Undergrowth protected the soil from erosion and leaching.[13]

Yet the very presence of this undergrowth, which was crucial to the reproduction of a bewar system, proved detrimental to the growth of a sal seedling. The dangerous effects of drought, forest fires and growth of thick bamboo and grasses were realised by the foresters as early as the late nineteenth century. Experiments in the sylviculture of sal showed that roots are weakened by the presence of undergrowth. As the *Indian Forester* argued, once the selections have been made for felling,

> the remaining periodic blocks will steadily close up during the first period, and it will be our business to do our utmost to bring about as dense a leaf canopy as possible, so as to kill out all the grass and render

[12] F. C. Osmaston, 'Sal and its Regeneration', *Indian Forester* (henceforth *IF*) 54: 12, December 1928, pp. 230–74 and pp. 639–54; see especially pp. 569–70.

[13] *Ibid*., p. 554.

> the forest practically fire proof, and to ameliorate the condition of the soil. When an area has once been thoroughly purged of grass by a long spell of cover, it is practicable, I believe, to somewhat slacken the density of the cover, as in such an area the grass can not immediately return.[14]

As the foresters saw it, the relationship between the density of the forest and the growth of sal was an adverse one. At any rate a 'good sal tree' would take about 15 years to reach its full height. In this way, the temporal cycle of sal regeneration, as the foresters wanted to practise it, and that of bewar conflicted with each other. There were also three other related points of contradiction between sal regeneration and the cultivation methods of bewar: the need for thick undergrowth and firing of the fields in bewar. The cultivation methods of the baigas consisted primarily of firing the field and burning the forest growth to manure the laterite soil. Thereafter, the seeds were broadcast in these ashes before the coming of the rains. Finally, experiments on *sal* showed that the strip method of planting would yield the highest production and the best results. In contrast, the swiddeners' choice of random sites (as the planters and foresters saw it) was incompatible with the organised method by which they sought to regenerate sal.

These conflicts between sal regeneration and bewar are not immediate because of the existence of mixed forests in the bewar area. Apart from sal, the other species found in the Maikals are described in Table 4.2.

Table 4.2: *Forest Trees and Grasses in the Maikals*

English (Latin) Names	*Local Name*	*Uses*
Sal (*Shorea robusta*)	Surye or Rinjal	The seeds are used for food and resin is used for medicines
Black Eyne (*Terminilia tomentosa*)	Saj	Used for breeding lac and its resin is sold to local craftsmen for dyes
Basewellia thurifera	Salai	Drought resistant tree used by Agarias to make charcoal

(Table 4.2 continued on next page)

14 Anon, 'The Most Suitable Treatment Applicable to Sal', *IF*, 24: 12, December 1898, p. 451.

(Table 4.2 continued)

Embilica officinalios	Aoula	Its fruit is eaten during famine
Grewia elastica	Dhamin	Used for making bows, arrows and traps
Bassia latifolia	Mahua	Used for liquor and food
Dichanthium	Khel	Principal grazing grass
caricosum. A	Sukra	Medicinal grass

Source: Drawn by Author.

In the western Maikals, sal grew with drought-resistant trees like *saj*, *salai*, and *aoula*. It was also surrounded by a variety of grasses like mosaic, *khel*—the principal grazing grass—and *sukra*.[15] John Forester, an official of the Indian Forest Service, enumerated the advantages of mixed forest growth as greater density and better retention of the soil. The forester argued that the natural reproduction of the mixed forests was easier than that of forests with a single species. Last, mixed forests were less prone to natural disasters than pure forests.[16] Mixed forests, thus, helped to neutralise the onslaughts of bewar on sal forests, and also helped to protect the forest communities from scarcities and famines.

The official sources of the nineteenth century reveal this conflict between the bewar and the natural regeneration of large *sal* trees. In the progress reports of the 1860s, foresters often referred to depletion of timber in the bewar areas. Both the contractors and baigas were blamed for the poor condition of sal in the low-lying areas of the Maikals. For example, Colonel Ward wrote that the baigas had 'destructive propensities' and their cultivation had destroyed the sal on the hillsides, laying the land absolutely bare. Similarly, Forsyth spoke of the baigas as the 'most terrible enemy of the forest', their biggest crime being the felling of large trees during bewar. As he noted, large sal trees were replaced by short stumps in bewar tracts.[17] Neither officials like Ward

[15] *Mandla District Gazetteer* (henceforth *Mandla DG*), 1909, Nagpur.

[16] John Nisbet, 'On Mixed Forests and their Advantages over Pure Forests', Appendix in *IF* 20, 1894, p. 5.

[17] H. C. E. Ward, 'Notes on Baigas,' in *Mandla Settlement Report*, 1869, p. 35. Also see the early inspection reports of Colonels Pearson and Hill in the *Forest Administration Reports* of 1864–90. This view is consistently upheld by the British; see also Forsyth, *Highlands of Central India*, p. 304.

and Pearson, nor travellers like Forsyth denied the existence of thick undergrowth and bamboo in the bewar areas. In 1939, Elwin selectively quoted from these sources to make a case for bewar. He argued that bewar cultivation had not led to the 'destruction of forests'. But the evidence that he provides is thin and focuses on the fast regeneration of undergrowth rather than trees. In fact, Elwin confirms Forsyth's opinion when he quotes J. B. Fuller's statements of 1898 and 1901:

> I am under the impression that in the past rather exaggerated ideas have been entertained of the injurious effects of *bewar*, especially in a country of heavy rainfall where its effects seem to be not of permanent denudation of the land, but of the substitution of one kind of forest growth for another…it is of much more importance that a tribe of people should live in peace and comfort, than a certain area should have trees of one sort or another or that trees should grow at all.[18]

The foresters of the nineteenth century objected to precisely this replacement of forests. By the 'destruction of forests' they implied the destruction of large sal trees that were considered valuable timber. The term 'forests' signified totally different things to Elwin. It connoted serene, harmonious and beautifully 'wild' surroundings inhabited by the forest communities. For this reason, Elwin was unable to counter the claim that banning of bewar was essential to the successful regeneration of timber forests. Contrary to Elwin's contention, the interests of bewar cultivation were often conflicting with the processes that led to the growth of large sal trees. In this situation, the baigas were left with no option but to practise bewar in mixed forests. This conclusion shows that issues related to the ecological viability of bewar were in fact more complex than Elwin's text portrays them to be.

The conflict between the regeneration of the sal tree and bewar unfolded the tensions within baiga society. Though anthropologists like Elwin contended that there was no contradiction between these two or that the identity and the survival of the baiga society depended solely on bewar, empirical evidence from the period suggests that this was not so. While bewar resulted in the distinctiveness of baiga life, gathering activities made this community a part of the wider local and non-local economy. In this sense, neither the forests nor the baigas were self-sufficient or isolated society in the mid-nineteenth century.

[18] Fuller quoted in Elwin, *The Baiga*, p. 129.

Local Production Networks and Baiga Livelihood

An examination of gathering activities in forests shows that baiga survival was intimately connected to their links with peasants, graziers and craftsmen who were important users of the forests. The baigas believed that the forest had everything they needed for sustenance. Since their primary source of survival was the forests and they were considered superior to the other forest communities because of their important ritual role, they saw themselves as kings of the forests. Their abhorrence of the plough, as we shall see later, only reinforced this self-image, bestowing priestly functions upon them during the sowing and harvest time.

This image was firmly grounded in the material basis of baiga survival. The forests were used by the baiga for gathering food, sal seeds and other minor forest produce. Gathering activities were undertaken by both women and men. Women gathered roots, fruits, leaves, mahua and sal seeds. Root and fruits were important sources of food in periods when bewar crop was ripening. Men also gathered roots in times of drought and famine. According to Russell and Hira Lal, a baiga could collect a full basket of roots in a single day.[19] Mahua and sal seeds were collected by women and used as food or for brewing *tadi* (country liquor). Apart from this, the collection of leaves was useful for household purposes and bamboo was collected to make baskets and mats that were sold in the markets.

Gathering was not a gender-specific task. Both men and women did gathering for the local agrarian market. This brought them in contact with people who did not live in the forests but depended in one way or another on them. The nature of the contact was, however, quite gendered in at least some respects. The more lucrative produce like honey, lac sticks, silk cocoons and tans were sold by the men to the craftsmen. In contrast, the women sold flowers, leaves and seeds of trees like the sal as well as fuelwood to peasants. These were presumably less lucrative than the produce sold by men. The domination of male members in controlling the most profitable ways of survival was also revealed in the cultivation system: the male members of the community

[19] Russell and Hira Lal, *Tribes and Castes*, vol. 2, p. 91.

controlled the technology and the means of production. For example, the women were not allowed to even touch the axe. The father was the head of the family, but the son could own his own axe after marriage. Once he got married, the father gifted an axe to his son. The gift of the axe was an important gesture that recognised the son as a co-equal producer. At the same time, it also created a hierarchy between households—the domination of the patriarch continued till the son attained that position himself. It is significant that the male child's access to the axe, bows and arrows gave him the potential to become a powerful member of the community. The same was not true of the women.[20]

The baiga patriarch reinforced his position vis-à-vis the women in their relationship with the peasants. Even as the baigas considered cultivation by the plough a sin, and most peasant castes as socially inferior, the structure of the regional political economy ensured that peasants were wealthier and had more social influence than the baigas. Apart from excluding the women from the decision-making process, the self-image of baiga men as protectors of mother earth had little effect on their social and economic position within the wider economy. They continued to eke out their existence from the precarious combination of swiddening and gathering. But by refusing to work as wage labourers, the men forced the women to do the bulk of the work and perform the tasks of transplantation and harvesting, especially during times of scarcity when bewar crops had failed. Men

[20] Elwin, *The Baiga*, pp. 78–79. Also see p. 273. In his text, Elwin argues that the baiga society is morally and socially superior to the modern one because of the freedom it provides to its women. As he wrote: 'In Baiga society, the women enjoyed an excellent position. Theoretically, in so patriarchal and priestly a tribe, man should be in the ascendant: actually, women have great freedom and no little authority. She may go about on her alone; she generally chooses her own husband, and changes him at will; she may dance in public; she may take her wares to a bazaar and open her own shop there, she may own property and she may drink and smoke in her husband's presence.' *Ibid.*, p. 275. But his evidence does not bear out this conclusion on the baiga women. In fact, the precarious survival strategies of the baigas were held together in a loosely structured system of survival by collective male authority that denied the women many rights on property and land. For this argument see Archana Prasad, 'The Political Ecology of Swidden Cultivation', *Tools and Tillage*, 7: 4, 1995, pp. 206–30.

only hired their labour out for one specific task: the uprooting of rice seedlings before transplantation.[21]

The fate of lower class gonds and baigas was sealed together by networks of commodity exchange. The forests saw a convergence of different occupational interests which supported the artisan and peasant economies. The mobility of the artisans between towns, villages and forests ensured the development of wider linkages of forests with other economies. The lac jewellery and toy makers were good examples of this. The craftsmen who worked with lac did not necessarily live inside the forest because lac propagation was regarded with certain superstitious repugnance by upper-caste peasants and craftsmen. Therefore, these craftsmen bought stick lac from the gonds, korkus and baigas.[22]

The lac seed swarmed twice a year, in October and June. Labour for its collection was available in June–July for the Baisakh crop, and in October–November for the Kartik crop. Thereafter, lac was taken to the markets where the forest-dwellers sold it to the craftsmen. The main lac market near Mandla was in Seoni where lac was in great demand.[23] After collection, stick lac was cleared of the wood and the encrusted twigs and barks were then placed in long cotton bags. These bags were put before the fire and heated, and the lacquer gum squeezed from the sticks and mixed with clay and other materials to make toys and bangles.[24] Since the best business for the *lakheras* was in the season of festivals, their income was seasonal and any transaction of the forest communities to propagate lac must have needed to take that into account. While their main markets were in small towns like Mandla and Seoni, they also sold bangles in the villages.

The presence of lac artisans in the suburbs of the towns and the mobility of the korkus, gonds and the baigas to sell the lac created wide-

[21] Russell and Hira Lal, *Tribes and Castes*, vol. 2, p. 91. Also see settlement reports of various districts for evidence of the employment of lower class gond and baiga labour by richer peasants of the same or other upper-caste communities.

[22] Russell and Hira Lal, *Tribes and Castes*, vol. 3, p. 105.

[23] Central Provinces Secretariat Records, Nagpur (henceforth CPSR), Forest Department, Case File No. 22 'B' Progs, April 1919, (henceforth Lac, 1919), p. 3. Also See CPSR, Forest Department, Compilation 50 (6), 1875 (henceforth Lac, 1875), pp. 85–86.

[24] Russell and Hira Lal, *Tribes and Castes*, vol. 3, p. 106.

ranging linkages between the urban, agricultural and forest societies. The forest communities found themselves at the lower end of these networks. They had to walk long distances to get to the markets and were compelled to sell their lac at prices on which the artisan agreed. This is so because the lakhera or artisans who moulded lac were superior to the baigas in the social hierarchy. The lakheras believed that their ancestors were *kayastha*s who had established friendly social relations with the Rajputs and *bania*s. In some societies they were ranked above the *kunbi*s. The kunbis considered the forest communities outcastes and it is likely that the lakheras treated them as subordinates.

Similar linkages between urban centres and forests were evident in the case of the dyers. The main dyers of the province were the *koshti*s, *chippa*s, *nillar*s, *rangari*s and *rangrez*. They dyed in traditional Indian colours, mainly reds and yellows, got from the roots of trees like *al*, *kusum* and *rohun*. Resin was extracted by tree tapping, and in some cases, by pounding roots. Forsyth does not record whether resin or gum was extracted by the dyer or bought from the forest-dweller. Rangrez and rangaris wove silk-bordered cloths and dyed their own thread with lac or *palas* flowers. Since they are said to have abhorred manual labour, it is likely that they bought stick lac and palas flowers from forest people. Apart from this, the *tussar* silk industry and the silk weavers were dependent on the collection of silk cocoons by gonds. Silk weavers (like koshtis) bought cocoons from the forest communities and boiled them, and wound the thread on reels. Dyers used myrabolan flowers to get blue or black dye, while red was made from lac. Other cotton weavers like chippas, nillars and the rangrez also bought flowers of a similar kind to make dyes for cloth.[25] The dyers did not necessarily develop close social relations with the forest communities. Those like the koshtis lived in towns and enjoyed a good income. The class difference gave them an economic advantage over the forest communities. But conversely, the hesitance of these artisans to go into the forest to collect their own resins also made them dependent on mobile forest dwellers.

[25] For this paragraph the material has been taken from Robert Vane Russell, *Monograph on Dyeing in Central Provinces* (Nagpur) 1910, Russell and Hira Lal, *Tribes and Castes*, vols 2 and 3, and C. E. Low, *Preliminary Notes on Cottage Industries in Central Provinces and Berar* (Bombay) 1908.

While the forest–township interface was characterised by the role of forests in the household production of artisans, the relationship of the forest with permanent cultivation was mediated by the use of forests by graziers like the *ahir*s and seasonal pastoral traders like the *banjara*s. Forest lands on the fringes of ryotwari villages were subsistence zones shared by peasants, forest dwellers and specialist herdsmen like the ahirs, *gowarie*s and the banjaras. The ahirs tended and grazed the cattle of individual peasant families in the forests.[26] Their intimate relationships with the gonds and others living in the forest were a result of these daily excursions. Russell notes that in Chanda, the ahirs had developed intimate links with the gonds. They not only employed gonds to graze cattle but also admitted them within the ahir caste. In fact, the origins of the gowarie community of Chanda were traced to the intermarriage between gonds and ahirs. In Mandla, ahirs were hardly considered Hindus as they lived in gond villages.[27]

The relationships of the banjaras with the peasant and forest communities followed a different pattern. An early article on the migratory people of India described the banjaras as grain merchants whose traffic was carried on the backs of bullocks. They traversed the most impracticable country for supplies which they delivered to districts where scarcity prevailed, and later moved to large army camps where they supplied grains. Thus, they established an important place for themselves in the political economy of the Central Provinces.[28] In British records, the banjaras were described as a caste of carriers and drivers of pack bullocks. They derived their subsistence from the sale of salt and oil to forest dwellers and the transportation of goods. Ethnographers have given instances of *charan*s and the banjaras carrying loads for the Rajput and the Maratha armies.[29] Banjara seasonal movement spanned a large territory. Their ability to negotiate thick forests and 'rugged'

[26] I am grateful to Sumit Guha for this observation.

[27] Russell and Hira Lal, *Tribes and Castes*, vol. 2, p. 24.

[28] Edward Balfour, 'On Migratory Tribes of the Natives in Central India', *Journal of the Asiatic Society of Bengal* 1: 1, 1844, pp. 2–3.

[29] For the banjaras see N. F. Cumberledge, *Monograph on the Banjara Clan* (Nagpur) 1882, rpt. Also see Col. Mackenzie, 'Notes on the Banjaras', in *Berar Census Report*, 1881, and Russell and Hira Lal, *Tribes and Castes*, vol. 2, pp. 163–92.

terrain made them indispensable to the rulers who relied on them to carry messages and arms. The British often condemned them as robbers and bandits due to this. Their utility to the peasant economy was minimal. Though they sold a few milk products to the peasants, they never grazed peasant cattle. This was done by more 'trusted' graziers and residents of the village, like ahirs and gowaries, who even though low caste, were certainly included in the system exchanges within the village community.

In Mandla, the main job of the *agarias* was to repair old sickles, axes and to supply new ones. For this they received an annual tribute of grain from the villagers called *jewar*. Then they got foodgrain in the month of Asad (known as *Bij Phutti*), and finally they got grains at the time of the harvest of both spring and autumn crops, known as *kainhari*. The agaria rates depended on the nature of the soil. Agarias were forced to find work on the trade routes to fulfil their needs when they were not repairing axes. At other times their survival was linked to the exchange of agricultural implements for grain. In the zamindaris of Bilaspur, all trade with the agarias took place through the medium of iron. Instances of exchange between the *marias* and the agarias were common in Bastar and Chanda.[30] The presence of the agarias incorporated the baigas into a wider agrarian economy. In times of famine, the agarias played an important part in sustaining the baiga economy. They sold them food, oil and salt, essential everyday baiga items.

There was a dialectical relationship between bewar and gathering. As the previous discussion has demonstrated, these economies were tied together in a loosely-structured survival system and had a precariously tenuous relationship. The artisan, pastoral and agaria networks linked the baiga with small townships as well as sedentary rural society. The mobility of these groups also revealed the critical, and often conflicting, relationship between the forests, agriculture, rural crafts and large-scale capitalist enterprise. The nature of baiga survival has to be seen in this context. Though in most cases the baiga formed the lower class in the political economy of the region, their actions had a profound influence over the livelihood of others, like the artisans. The presence of the baigas

[30] See *Mandla DG*, pp. 179–80. Also see the Settlement Reports of the Bilaspur zamindaris.

in the agrarian economy created an antagonistic contradiction within the internal structure of their own society. Bewar required the lopping of large sal trees that were prevalent in the forests of the Maikal ranges. It also interfered with the natural regeneration of sal forests. At the same time, the survival of the baigas was increasingly dependent upon the existence of such trees. This was particularly true of gathering of sal seeds, *ral* and branches as important items of sale in the local and regional economy.

In this sense baiga subsistence witnessed two tendencies. While bewar was increasingly dependent on local networks for its survival, the community identity made a strong distinction between bewar and other activities. Bewar was considered a sacred activity by the patriarchs of the baiga society. A myth recorded by Russell and Hira Lal in the *Ain-i-Akbari* shows that the importance of bewar in baiga identity and livelihood existed in the medieval ages:

> In the beginning they say, God created Nanga Baiga and Nanga Baigin, first of the human race, and asked them by what calling they would choose to live. They at once said they would make their living by living in the jungle, and permission being accorded, have done so ever since. They had two sons, one of whom remained a Baiga, while the other became a Gond and a tiller of the soil. The sons married their own two sisters...and while the elder couple are the ancestors of the Baigas, from the younger are descended the Gonds and all the remainder of the human race. In another version of the story, the first Baiga cut down two thousand old *sal* trees in one day, and God told him to sprinkle a few grains of *kutki* on the ashes, and then retire and sleep for some months, when on his return he would be able to harvest a rich harvest for their children.[31]

According to this myth, the baigas were the ancestors of all the other races. Their practice of bewar and dependence on forests was considered a boon because of their superior status. Because of this the baigas looked down upon the gonds and other forest people as socially inferior. Baigas considered themselves as different, and the continuous replication of this myth only furthered the segregating trend in the bewar system.

[31] Russell and Hira Lal, *Tribes and Castes*, vol. 2, p. 79.

Baiga *Chak* and Exclusion of Bewar from Local Networks

The pattern of baiga survival, established after a considerable amount of upheaval in the Gondwana, could only remain in a precarious balance under limited physical and demographic conditions. Until the middle of the nineteenth century, the forests around the baiga *chak* (reserve) were extensive and pressure on forest land was limited. However, with the implementation of the colonial agrarian policy the pressure on these forest and pasture lands increased significantly and, therefore, the tensions between gathering and bewar activities were greatly accentuated. In this situation, the Nanga Baiga myth, too, underwent a change. It acted as a catalyst in the segregation and the isolation of the baiga people from the rest of the economy: a process that was the inevitable result of the long dialectical interaction between the tensions in the baiga economy and the colonial administration.

Nineteenth-century perceptions of bewar cultivation were marred by confused classifications of this form of cultivation. These perceptions can be identified in early forest and revenue surveys. Swiddening by the baigas was classified either as dhya, *dhaiya* or bewar. In the forest report of 1862–63, Richard Temple wrote:

> One great cause of the wastage and destruction of the forest is called *Dhya* cultivation.... This *Dhya* cultivation is practically a substitute for ploughing and a device for saving trouble of that operation. It is resorted to by the hill people who are averse to labour and have practically no agricultural capital.[32]

Temple employed the standards of a capitalist agrarian economy and classified the dhaiya economy as 'primitive' or 'backward'. Its traits were laziness and wastefulness, not acceptable behavioural patterns of capitalist societies. The most significant feature of this perception was that it was not based on a topographical or ecological survey of the region, but on the strong conviction that it harmed sal forests. For example, Forsyth wrote:

[32] *Report of Forest Administration, Central Provinces*, 1863–64, Calcutta, p. 84.

> The Byga is the most terrible enemy to the forests we have anywhere in these hills. Thousands of square miles of *sal* forests have been destroyed by them in the progress of their *dhya* cultivation, the ground being afterwards occupied by dense scrub of low *sal* bushes springing from the stumps. In addition to this, the largest trees everywhere have been girdled by them to allow the gum resin of the *sal* to exude.[33]

As we have seen, such opinions were not entirely wrong. The argument was formulated from the standpoint of the official need to preserve large sal trees. In this context, the first settlement report of Mandla pointed out the need to control dhaiya. By the mid-1870s, all officers of Eastern Central Provinces started linking efforts to control bewar with forest conservancy. For example, for the Chief Commissioner of the Provinces, Mr Croswaithe,

> There are only two courses open to the Baiga: either he be induced to give up *bewar* and take to plough cultivation or we make it impossible for them to practise *bewar* in these valuable jungles.[34]

This view was expressed by Croswaithe at a time when the reservation of forests was already in place by the Forest Act of 1865. The fact that the statement was made before the Forest Act of 1878 only accentuated its significance. This act detailed the rights of the local population in forests. However, by doing so, it also restricted the mobility of the local population in the forests. It was this feature of forest laws that conflicted the most with the baiga lifestyle.

Movement from one field to another was the basis of the success of bewar and was necessitated by the long fallow cycles. It led to the control of large forest tracts over a long period of time. When the baigas moved from one field to another, their huts remained in the village. Thus, they controlled at least two or three fields at a time. This contradicted the British need to have permanent and absolute territorial control to achieve the aims of forest conservancy and the expansion of plough cultivation. So adamant were the colonists to take over territorial control that they wished away the natural constraints that

[33] Forsyth, *Highlands of Central India*, pp. 304–5.

[34] Madhya Pradesh Secretariat Records, Bhopal, Forest Compilation No. 178, 1878, p. 18.

affected the practise of plough cultivation in thick tropical forest conditions. Thus, Forsyth opined:

> As for the obstacles supposed to be presented by the rank vegetation and noxious animals, they are chiefly imaginary. Immense plains lie ready for the plough, if merely the coarse natural grasses were cleared away, there being no brushwood or heavy timber to speak of. The luxuriance of grasses is only evidence of the fatness of the land that lies below; and a torch applied in the month of May will, over large tracts, remove all obstacle to the immediate application of the plough.[35]

Like many other officials of the late nineteenth and early twentieth century, Forsyth believed that forest communities used the axe for social and cultural reasons. Some administrators also felt that they wanted to use the plough but did not have the requisite knowledge or means to do so. In 1869, the settlement officer of Chanda reported the marias wanted to use the plough but did not know how to operate it. The settlement officer highlighted the nexus between the moneylender who lent the plough and the maria desire to use it. He thus expressed the need for the 'improvement and the protection' of marias.[36]

The colonial project of protection was integrally connected to forest conservation, its main aim being to curtail the mobility of forest people. The first serious attempts to 'improve' baiga agricultural methods were made after Ward's note of 1868–69. Its main thrust was to wean the baigas away from the axe. He also wanted to help them give up their migratory habits so that they would settle down like ordinary peasants and make their contribution to government revenues. Ward reported:

> I had many a talk with the village panchayat as regards the manner of their cultivation, and was rather put in to reply to some of the pertinent questions of the shrewd old men occasionally. My efforts to persuade them to take to the plough were quite unsuccessful and after hard pleading I could only induce them to confine their depredations on the forest and take up land accorded to them in regular rotation.[37]

[35] Forsyth, *Highlands of Central India*, p. 309.

[36] Annual Report of the Forest Department, Central Provinces, 1862–63.

[37] Ward, 'Notes on Baigas', p. 160.

Ward's early efforts showed that the British Empire had to contend with the power of the local institutions in order to regulate individual behaviour. In doing so, the interaction between these institutions (manned by village elders and the priests), and colonialism also led to the reproduction of an aggressive baiga identity. Although the creation myth also privileged bewar over other survival practices, the aggressiveness of bewar identity was not witnessed till the 1860s when baiga elders asserted that 'they were "Baigas", their fathers had been born "Baigas" before them; that they had for generations cut *bewars* and would continue to so. Let the "sirkar" do whatever it pleases to try and prevent it.'[38]

This response was rooted in the baiga self-image that they had a birthright to cut bewar. The stance of the baiga elders had two broad effects. In the first place it led to the beginning of the resolution of the contradiction of within the early nineteenth-century baiga economy for it created the basis of the segregation of the bewar-practising baigas into the baiga chak. Second, the uncompromising stance of the baiga elders and their threat to break the law prompted the government to negotiate with them.

The Chief Commissioner of the Provinces, Croswaithe, took a hard line on the issue. He ordered that the baigas should not be allowed to take 'the law into their hands'. Other officials like the Deputy Commissioner of Mandla, Major General Miller, favoured the establishment of a separate baiga settlement. The baigas would be induced to take up the plough within this settlement. Even if the effort to socialise the baigas into sedentary cultivation failed, the British hoped that their bewar cutting could be controlled in order to meet the aims of conservation. In July 1878 Miller reported that 200 baiga families were willing to settle under such a scheme.[39] In 1890 two villages and four *tolas* or settlements of houses were chosen to form a 'Baiga reserve'. Originally comprising 362 families, these villages were in the Dindori and Karanjiya forest reserves. Tracts were to be allotted only to baigas and were to be available for further colonisation by them. Miller's position was a compromise between the baiga refusal to take

[38] CPSR, Forest Department, Compilation 178, 1878, p. 47.

[39] *Ibid.*

up the plough and the determination of the obstinate colonists to ban bewar. In this sense, change occurred through the interactions of the district administration with baiga patriarchal authority.

In Mandla, the will to compromise was adopted after experiments to socialise the baigas to plough cultivation were taken up in Balaghat district. After the initial failure to induce the baigas to take up the plough, the dangers of sudden changes were realised. In 1867–68 the Conservator reported that he was:

> more convinced than ever of the impossibility of putting an immediate stop to this practice forcibly: it would not only be of great hardship to the Baigas who would be reduced to starvation but it would possibly drive a peaceful and orderly community to resort to crime as a means of livelihood.[40]

The foresters of the Province realised that the livelihood of the baigas needed to be protected if they were to be induced to cooperate with the foresters. The officials hoped that the baigas would provide labour for the departmental forest works if they were allowed to practise bewar in a limited area. For this reason the first efforts to curtail bewar were made in Balaghat in 1878. Bewar was to be allowed in a few chosen villages in the Baihar *tahsil* of this district. These villages were within the limits of the Reserved Forest. The Commissioner of Balaghat reported that 18 baiga families had given up bewar cultivation and settled with gonds in the upland villages. Through this process, the government hoped to introduce the baigas to a civilised life and the use of the plough. For this the government was willing to provide the baigas cattle, seeds and a plough in every village.[41] Early results were reported by Commissioner Newmarch:

> These families are not recognisable from respectable Gonds and the headmen viz. Mubud and Runjar are well-dressed men as the best Gonds I have seen. Besides the above several patels in Beihar uplands at the instance of the tehsildar, have reclaimed individual Bygas and

[40] Annual Review of the Forest Department, Central Provinces, 1867–68, p. 9.

[41] Ward, 'Notes on Baigas', p. 160.

> settled them as cultivators...*Dhya* cultivation is exclusive to some government protected villages and the zamindaris.[42]

The baigas had obliged the government by making attempts to give up cultivation of bewar and pursue other subsistence activities like rope-making, fruit gathering and labouring for the gonds. Provoked by this, the administrators evolved more stringent and exacting measures to keep them away from the axe. In the same year, the Commissioner issued instructions under which the baigas were to be settled on leased land and new clearings were to be discouraged. Bewar crops were to be destroyed if the baigas were caught doing bewar cultivation. In addition, all cultivators were to pay a rent of Rs 2 for every plough with forest and grazing dues. The government, however, retained the right to enhance the rent by Rs 4. Each cultivated strip was equivalent to one plough.[43] Of these measures, the step of burning the bewar crop was the harshest and was prompted by the partial success of the Balaghat policy. In the mid-1870s, Colonel Fischer, the District Commissioner of Balaghat, contended that 'the Baigas have largely improved the face off Balaghat... They have migrated to Kawardha or Sakh Tehri and only about 250 families have been left in Balaghat district. Of these fifteen or twenty families have settled down to plough cultivation'.[44]

The Balaghat administration had overestimated the early impact of the policy to curtail bewar in Balaghat, and any attempt to apply coercive measures was met by desperate reappraisals. As the Commissioner of Nagpur pointed out, the baigas would either starve or die. By 1880, the scheme faced certain reverses. All the village elders appealed to the government to lift restrictions on bewars.[45]

The process of negotiation between the baiga elders and the local colonial administration had brought about changes in the management of land. The baiga priests and elders had lost control over a large area of forest land by the mid-nineteenth century when the first forest laws were implemented. The baiga institutions did not play an important role in the distribution of land to individual families. Nor was the labour process based on the interpretation of the old baiga myth. The seasonal rhythm

[42] CPSR, Forest Department, Compilation 178, 1878, pp. 4–5.

[43] *Ibid.*, pp. 47–48.

[44] *Ibid.*, p. 36.

[45] CPSR, Forest Department, Compilation 178, 1881, p. 7.

of survival was broken by the very act of reservation. In the 1880s, the debate within the officialdom brought the process of internal change within the baiga community in sharper focus. Community institutions could no longer control the nature of cultivation as the colonists started discussing the best possible ways to expand plough cultivation in order to stop axe cultivation. In the 1880s, the Conservator of Forests wrote:

> As long as permanent cultivation is confined to the flat country, everybody shall benefit from it. Sufficient thought has not been given to the extension of cultivation in suitable areas and that the extension has often led to the abandoning of old villages and fields for new ones.[46]

The Conservator's comment was the first in a series of comments that realised that plough cultivation was not the most viable form of cultivation in the Maikals. He implied that the officials promoting the introduction of the plough in these areas had not thought carefully about the ecological assumptions of that measure. He further stated that the abandonment of fields was not merely a form of protest but also a way of avoiding a decline in productivity. The schemes to exchange forest lands held by the baigas for cultivable lands also failed miserably. The plan was abandoned since the baigas refused to move from the forest. Foresters also asserted that it was unlikely that a large section of the baigas would ever turn to settled agriculture. Further, the government policy of granting land rights only to those baigas who had been cultivating the same piece of land for 30 years would leave most others destitute. But this view was coupled with the foresters' desire to control bewar. The fight between the forest and the revenue departments for the control over land was to continue with greater vigour in the early twentieth century.

The harsh measures taken by the British government to introduce the plough had led to indebtedness and destitution of at least 485 baiga families. Major Bloomfield pointed out that only 75 baiga families had received any kind of government loans or assistance to survive and about 40 families had been able to fend for themselves. The rest were living either in debt or abject poverty.[47] While the colonial administration made plans for alternative livelihood for the baigas, their attempts at

[46] *Ibid.*, p. 3.

[47] *Ibid.*, pp. 11–16 and Compilation 178, 1880, p. 61.

stopping bewar were not as successful as they had hoped. By 1881, officials reported that most baigas had attained the status of squatters on their bewar lands. In fact, Major Bloomfield's report on the baigas in 1881 admitted that the produce of bewar cultivation coupled with gathering and sale of minor forest products had helped the baigas to survive so far. He also suggested that if the baigas could not take to regular permanent cultivation, alternative forms of livelihood should be created for them so that they could support themselves. He felt that labour opportunities in forest works should be protected for the baigas and they should be given rights in fisheries. The *tahsildar* of Baihar suggested other occupations for the baigas: village *pujari*ship (priesthood), bamboo rope and basket-making, hunting, labour and selling forest produce.[48]

The initial attempts to control bewar were quite unsuccessful. By the turn of the century, the aims of the forest department changed from a policy of complete eradication of bewar to one of containing it to a limited area. The baiga chak was formed in 1890 with this objective in mind. In the chak, 22,858 acres of land were reserved for bewar cultivation by the baigas. The chak consisted of five villages and nine clusters under the jurisdiction of the District Collector and the tahsildar.[49] There is little information about how many bewars were granted during this period. But there is a record of the baiga protests against the insufficient allocation of land for their bewars and the displacement that they suffered due to the formation of the chak. Dharmu Baiga of *Mouza* Oudhar wrote to the tahsildar about the 60 baiga families that were to be removed to Dhurukota:

> I am residing in Mouza Oudhar and I am cutting *bewar* in this village for the past three years. Sixty families of the Baigas inhabit these villages and they all live upon *dhya* cultivation. We have now been ordered to leave this village and remove to Dhurukota....It has very little jungle and most of it was cut down by old Baiga settlers; and consequently it cannot support other Baigas of different villages.... We therefore pray that we may be allowed *bewar* cutting in adjoining villages.[50]

[48] *Ibid.*, p. 22. Also see Compilation 178, 1878, pp. 63–79.

[49] Mandla District Records, Mandla Collectorate, Mandla, Forests File No. 37, 1937, p. 105.

[50] CPSR, Forest Department, Compilation 178, 1890–92, pp. 11–12.

The statement shows that most baigas living outside the chak were not willing to move into the chak leaving their homesteads. Dharmu Baiga and the people that he represented lived outside the chak and refused to move to Dhurukota, a village in the centre of the chak. Instead, they begged to be allowed to cut bewar in a limited area around their own or neighbouring villages. The Conservator of Forests for the Northern Circle, Mr R. H. E. Thompson noted that all baiga families, except those who had willingly taken to plough cultivation, were to be deported into the chak. In the beginning, the reservation of forests was used as an excuse to move people. By 1891–92 it was reported that in Ramgarh tahsil only 80 families were living outside the reserve. In these 80 villages permanent cultivation was taken up willingly. For those baigas who did not either want to go to the chak or take up the plough, the government decided that four villages would be set up at Karadih, Udhan, Pandpur and Duldul for the supply of labour to the forest department. These villages were not to be solely baiga villages but gonds and agarias were also to be settled here.[51] In 1895, gonds and agarias were allowed to take up land inside the chak, furthering the policy of acculturation that was first started in Balaghat.

The process of negotiation and the formation of the baiga chak led to the consolidation of a new *Nange Baiga* myth in the late nineteenth and early twentieth centuries. Elwin recorded the following version of the myth in the initial decades of the twentieth century:

> The Baigas were established in the practice of *bewar* by Bhagwan himself, who when he called the tribes of the world to make a king first chose the Baiga. But *Nanga Baiga* begged that the Gond, his brother, be made king instead. Bhagwan was pleased, and as a mark of favour, took, *Nanga Baiga* by the hand and placed him on his throne by his side. He granted his prayer to make the Gond King, but gave him a greater blessing.
>
> "All kingdoms of the world" he said "may fall to pieces but, he who is made of the earth and is Bhumiraja, Lord of the Earth shall never forsake it. You will make your living form the Earth. You will dig roots and eat them. You will cut wood and carry it on your shoulders. Your wife will pick leaves and sell them. You must not tear the breast of your Mother, the Earth, with the plough like the Gond

[51] *Ibid.*, pp. 22–23.

> and the Hindu. You will cut down trees and burn them and sow the seeds in their ashes. But never become rich, for if you did you would forsake the earth and there will be no one to guard it and keep its nail in place".[52]

By projecting their identities as guardians of Mother Earth, the baigas countered the charge of officials like Ward and Temple. Baiga men asserted that they occupied a superior position in the agrarian society because of their abhorrence of the plough and their ability to abstain from all objects of wealth. And in order to maintain this superiority they would not practise any other form of survival. In this way they aided their own marginalisation. The assertion of this identity also led to a new contradiction within the baiga society, that between their material and the cultural practice. By making bewar the central focus of their identity they ensured that the colonists created a 'distinct space' for them within the agrarian society. Consequently, the potential conflict between the gathering and the bewar economies was accentuated, making the baigas indirect participants in their own impoverishment and segregation.

Conclusion

This chapter has sought to understand the ways in which the survival practices and the livelihood systems of the forest communities can be related to their local and the regional political economies. I have argued that the importance of bewar practices in baiga life has to be seen in the light of other production networks. The relations of production between the baigas and other users of the forests allowed the survival of the baigas during the long fallow periods in the cycle of bewar cultivation. And since these fallows were essential for the stability of bewar, this form of cultivation was intrinsically dependent on the baiga's links with their local and political economy.

The integration with the local economy was opposed by the self-image of the baigas. Bewar was always accorded a privileged position in baiga mythology. The *Nanga Baiga* myth of the late medieval ages

[52] Elwin, *The Baiga*, pp. 106–07.

emphasised the difference between bewar and other forms of survival. Bewar made the community different from others and emphasised the role of the baiga patriarchs as the kings of the forest. This enhanced their social status within their own community as also vis-à-vis other forest communities.

The self-image promoted by the *Nanga Baiga* myth of the medieval ages accentuated the main contradiction within nineteenth-century baiga society. On the one hand, this article has shown that the baigas were an intrinsic part of local and regional networks; on the other hand, I have argued that bewar practices also tended to segregate them from their local context. The conflict between the bewar cycle and the natural regeneration of sal forests played an important role in the intensification of the internal conflict between segregation and integration in baiga society.

The role of scientific forestry and the formation of the baiga chak in the history of the baigas needs to be considered in this context. The formation of the baiga chak can be considered to be a compromise between the colonial aims and the contingencies of baiga survival. The formation of the chak resulted in the initiation of two processes that resulted in the marginalisation of the baiga community since the 1930s. First, it cut the baigas off from the local and regional political economy. Second, it brought a change in the nature of the *Nanga Baiga* myth in the early twentieth century and this change was mostly a result of the consolidation of baiga identity around bewar. The aggressive articulation of the baiga identity as primarily bewar cutters aided in the segregation of the baigas from their local context. It enabled the colonists to demarcate an area for bewar and in the process pretend that they had taken care of the survival of the baigas. By the same measure, the emphasis of powerful baigas on the right to cut bewar at the expense of other rights in forests also aided in the segregation of the baigas from the local society, the *Nanga Baiga* myth becoming a catalyst in this process. Thus, Elwin's contention that the material and psychological well-being of the baigas was necessarily linked to their right to do bewar cultivation is problematic.

The Twilight World of the Caravan, the Regulated Market Economy and the Caravanners in Hyderabad State*

Bhangya Bhukya

Before the advent of the modern market and transport system in India, the *lambadas* were known as major caravan traders and merchants. The caravan trade dates back to ancient times. From the eleventh century, as trade and commerce expanded, it began to flourish, peaking during the Mughal period, as the imperial army maintained and employed huge numbers of nomadic groups as baggage carriers and foodgrain transporters.[1] Several nomadic and pastoral communities were involved in this trade, the lambadas being a principal one among them, particularly in south India. Their caravan trade networks were highly organised and played a crucial role in the growth of large-scale long-distance trade from the sixteenth century. Their role was particularly important during famines, as foodgrain was supplied from surplus of deficit areas through their caravan network. In addition, they also served as commissariat foodgrain and baggage transporters to armies during wars. They served the Sultanate, Mughal, French, British, and Nizam armies as independent transporters. They also acted as carriers for other merchants and dealers, and were often merchants in their own right as well, a fact that suggests that at least some had accumulated capital.[2] However, with the emergence of the regulated market and new economic relations enforced by the colonial

* Bhangya Bhukya, 'The Twilight World of the Caravan: The Regulated Market Economy and the Caravanners', in Bhangya Bhukya, *Subjugated Nomads: The Lambadas under the Rule of the Nizams*, Hyderabad: Orient BlackSwan, 2010, pp. 27–63.

[1] C. A. Bayly, *Rulers, Townsmen, and Bazaars: North Indian Society in the Age of British Expansion*, New Delhi: Oxford University Press, 1983, pp. 28–29.

[2] Edward Balfour, ed., *On the Ethnology of Hyderabad in the Dekhan*, Madras: Adelphi Press, 1871, pp. 12–13.

rulers, the caravan traders were marginalised.[3] This, coupled with the introduction of the railways and the growth of infrastructure, ruined the lambadas' caravan trade. In fact the lambadas suffered harsh treatment at the hands of the colonial state. Their caravans had been used by the colonial powers when the latter were first establishing themselves and expanding their markets. Once they stabilised their position, the lambadas fell out of favour. Now, their nomadic ways appeared to inhibit the formation of the new market economy and the modern state.[4]

The colonial state did not view the long-established trade and market practices of groups such as the lambadas positively, considering them to be backward and irrational.[5] Fairs and *jataras* (pilgrimages), at which communities converged for trade and transaction, were discouraged and suppressed by the colonial state through new regulations and customs, and transit duties. These practices inhibited the local market and exchange relations. The colonial state, indeed, appropriated and refashioned long-established market places by erecting regulated markets in towns and urban centres or fixed market places.[6]

In this essay, we seek to examine how the colonial regulation of the market economy combined with the introduction of railways and the growth of roads ruined the lambadas' caravan trade in Hyderabad State. This study is not meant to examine the economic history of the region, but it primarily focuses on the social history of the evolution and decline of the lambadas' caravan trade. The following questions will be dealt with. How did the nomadic lives of the lambadas drive them towards the caravan trade? How was the lambada community formed from the various subaltern communities? How did the lambadas' history

[3] Asiya Siddiqi, 'Introduction', in Asiya Siddiqi, ed., *Trade and Finance in Colonial India, 1750–1860*, New Delhi: Oxford University Press, 1995, p. 43.

[4] National Archives of India, Hyderabad Residency Records (Henceforth NAI, HRR), vol. 568, part II, 1782–1888, p. 378.

[5] K. N. Chaudhuri, *Trade and Civilisation in the Indian Ocean: An Economic History from the Rise of Islam to 1750*, Cambridge: Cambridge University Press, 1985, p. 222.

[6] Sudipta Sen has brought out how the colonial establishment considered traditional fairs and jataras in Bengal as obstacles to its business and destroyed them ruthlessly through various regulations. See his *Empire of Free Trade: The East India Company and Making of the Colonial Market Places*, Philadelphia: University of Pennsylvania Press, 1998, pp. 1–18.

of subalternity relate with the larger agenda of the colonial state? How did the colonial state suppress the local long-standing market practices and relations? How did the colonial establishment create dominant market forces to meet its ends? And how did these forces scuttle the lambadas' trading opportunities? While addressing these questions, I aim to also delineate the role played by the lambadas in transporting foodgrains during times of distress, and in establishing a foodgrain market network in rural Hyderabad. These questions help us situate the lambadas' history within the colonial market economy. Additionally, this essay seeks to understand the various other problems encountered by the lambadas in the face of colonial rule.

Before we take up the questions mentioned above it would be helpful to look at the history of the origins of the lambadas. Particularly, we need to try to understand how the lambada community was involved historically with the caravan trade, how the colonial ethnography and state endeavoured to create a homogeneous community of caravan traders; and how the lambadas came to south India in search of a livelihood.

A Brief History of Lambada Caravanners

The history of the lambadas begins with the caravan trade. Communities of nomadic caravanners were found all over the subcontinent, each with its own distinctive historical, cultural and social background. They appear to have been loose groupings, assimilating members from others involved in such a trade or even from settled peoples. Being a flourishing trade, there were attractions in taking up such an occupation. In the course of time, each group developed a varied social and racial composition and past. The communities of nomadic caravanners have different names and social roots in every region, and they claim a variety of statuses within the Hindu caste-society.[7] Their complex pasts and the scanty record of their early history make it difficult to understand the social history of these groups.

It is only from the sixteenth century that we find fairly detailed descriptions of them in the accounts and diaries of European travellers,

[7] R. E. Enthoven, ed., *The Tribes and Castes of Bombay*, 2 vols, Bombay: The Government Central Press, 1920, II: pp. 331–35.

and later on in the notes of colonial ethnographers and anthropologists. These groups especially attracted the attention of travellers and colonial ethnographers, because they traversed the country as grain suppliers to the army and came to be categorised as a 'criminal community'. However, these accounts and studies offer a fragmented and distorted view of the groups' histories. The colonial state had commissioned a number of studies on all the caravan groups in order to club them into one homogeneous community. These studies classified many diverse groups under the *banjara* community, which included the lambadas. Although it is unclear as to what the classification criteria employed were, primarily the following three were used: (i) those who had a Rajasthani ancestry and whose language was akin to Marwari; (ii) those whose means of livelihood were in some ways connected with oxen; and finally, (iii) those who were involved in the transport of grains and other items.[8]

There was a debate about the origin of the term 'banjara' among nineteenth-century ethnographers. One group argued that the term was derived from the Persian *biranjar* or rice-carriers, while another believed that its origin could be traced to Sanskrit words like *vanjari van* (forest) and *char* (to wander)—meaning forest wanderers, and was eventually spelt as banjara.[9] H. M. Elliot, a leading nineteenth-century ethnographer, who standardised the term 'Banjara', had a different viewpoint. Dismissing the earlier arguments, Elliot felt that the term originated from the story of Pramati in the *Das Kumara Cheritra*, a play written by Dandi, a predecessor of Kalidasa in the fifth century.[10] In the play, Pramati mentions a cockfight in the banjara camp, and he suggests that the term was derived from the Sanskrit *Vanijya* or *Banijya-kara*, a merchant. Elliot sought to prove his point using Mageshtanish's *Indica xi* (fourth century BCE), in which banjaras were an ascribed social class; thereby arguing that the term is of Indian origin.[11] It is apparent that

[8] Robert Gabriel Varady, 'North Indian Banjaras: Their Evolution as Transporters', *South Asia* 2: 1–2, 1979, p. 2.

[9] Enthoven, *The Tribes and Castes of Bombay*, II, p. 331.

[10] There is uncertainty about Dandi's period, but it is generally considered that he belonged to the eleventh or twelfth century.

[11] H. M. Elliot, *Memoirs on the History, Folk-Lore and Distribution of the Races of the North Western Provinces of India*, ed. John Beames, London: Trubner and Co., 1869, I: pp. 53–54.

Elliot's constructions were drawn from Orientalist racial theories, which argued that non-Aryan language speakers were the oldest races of the sub-continent.[12] William Crooke carried out a similar exercise at the end of the nineteenth century and he agreed with Elliot.[13] However, we have no significant evidence that suggests a relation between the ancient banjaras and the modern lambadas. Nevertheless, this term was standardised to describe all caravan communities, and it was added to the *Supplement to the Glossary of Indian Terms* compiled by H. M. Elliot in 1842.[14]

Let us now examine the following: when did the communities of nomadic caravanners first appear in the historical accounts? What are the terms used by medieval chroniclers and early European travellers to describe the nomadic caravanners? How were those terms replaced with the modern term 'banjara', when their accounts were translated into English or edited by nineteenth- and twentieth-century ethnographers? These questions would help us to understand the mentality of the colonial anthropologist.

Historically, Zia Barani's account of Ala-ud-Din Khilji's (c. 1296–1316) market-price control measures provides the first description of such a community, using the term *karwanis* (caravans) to describe them.[15] A later account by Shaikh Nasiruddin (c. 1355) describes such people as *nayaks*.[16] The term banjara first appears in Kabir's allegorical

[12] A similar exercise had also been done by George Abraham Grierson. He identified nomadic communities' (gypsies') languages with that of the Bhili, one of the oldest languages spoken in the Khandesh region of western India, because, according to him, the banjaras belonged to the oldest race—that is, Dravidian. See his *Linguistic Survey of India: Gipsy Languages*, Calcutta: Superintendent of Government Printing, 1922, XI, pp. 10–11; Susan Bayly, 'Caste and Race in the Colonial Ethnography of India', in Peter Robb, ed., *The Concept of Race in South Asia*, New Delhi: Oxford University Press, 1995, p. 172.

[13] William Crooke, ed., *The Tribes and Castes of the North Western Provinces and Oudh*, 4 vols., Calcutta: Superintendent of Government Printing, 1896, I, pp. 149–150.

[14] Varady, 'North Indian Banjaras', p. 12.

[15] Zia Barani, *Tarikh-i-Firuz Shahi*, ed. Saiyed Ahmad Khan, Calcutta: Bibliotheca Indica, 1862, pp. 305–07.

[16] Cited in Irfan Habib, 'Merchant Communities in Precolonial India', in James D. Tracy, ed., *The Rise of Merchant Empires: Long-Distance Trade in*

verse (c. 1500) as well as in Guru Nanak's.[17] Badauni, who came to Akbar's court in 1574, used the term banjaras, meaning grain-sellers, and he illustrated their importance in this capacity by giving examples, especially during the Mughal invasion of Mewar and Bengal.[18] Thomas Roe, East India Company's ambassador to India From 1615–19, has also described them at length, using the term 'caravan people.'[19] However, the term 'banjara' became popular by the beginning of the nineteenth century and it was reproduced in all medieval accounts translated into English to designate caravan traders. For instance, Mahomed Kasim Ferishta, a Persian historian who came to Bijapur in 1589, described them as Hindoo Bunjaras.[20] It was John Briggs, who edited and reprinted Ferishta's account in 1829, who modified this term as banjara. A similar modification is also found in all edited volumes of European travellers' accounts. We find a long and interesting description of caravan people's lives and manners from Tavernier, a seventeenth-century French traveller. He describes them as *manaris* and classifies them into four groups based on the goods they traded. According to his classification the first group traded in corn, the second in rice, the third in pulses, and the fourth in salt. In addition to trading, each group owned about one hundred thousand herds. But William Crooke, editor of *Travels in India*, supposes that manaris was probably a corruption of the term 'banjaras', and merges the four groups into one big community, dismissing the divergent cultural traits of the four.[21] The colonial ethnography/anthropology thus clubbed all the caravan bodies into one social group.

---

the Early Modern World 1350–1750, Cambridge: Cambridge University Press, 1990, pp. 371–99; chiefs or head men of banjara groups were called nayaks.

[17] *Ibid*., p. 374.

[18] Al Badaoni, *Muntakhabu-t-Tawarikh*, trans. and ed. George S. A. Ranking, 3 vols., New Delhi: Renaissance Publishing House, 1986, I, p. 477; II, pp. 182, 240.

[19] William Foster, ed., *The Embassy of Sir Thomas Roe to India, 1615–19*, London: Oxford University Press, 1926, p. 313.

[20] M. K. Ferishta, *History of the Rise of the Mahomedan Power in India till the Year 1612*, trans. John Briggs, 3 vols. repr., New Delhi: Oriental Books Reprint Corporation, 1981, II, p. 243.

[21] Jean Baptiste Tavernier, *Travels in India*, ed. William Crooke, London: Oxford University Press, 1925, I: p. 34.

The grouping of these divergent communities into one grand social stock of homogeneous community was, indeed, tempered by the colonial project of knowledge creation as well as by the European racial theories of the nineteenth century. The colonial state collected social and demographic data of the colony for multiple usages. It conceived its anthropological knowledge of the colony as the principal device to understand and control its subjects and to legitimise its hegemony.[22] The colonial state sought to implement this project of knowledge creation more rigorously from the middle of the nineteenth century. It was mainly because of the racial characterisation of the 1857 revolt that seriously challenged the British power in India, as well as the emergence of a more aggressive racial theory in western Europe. In this shift, every caste began to be defined in terms of race. This eventually resulted in opening up interesting discourses on caste and the caste system as a whole, and eventually occupation emerged as one of the main criteria in defining caste. From these discourses, two main schools of thought emerged: one led by Risley, which compared caste with race; and the other by Ibbeston, which endeavoured to see caste as the fundamental organ of the Indian social system.[23] In the identification of the races of India, indeed, the subaltern communities received greater attention, as the racial theorist believed that these were the oldest races and still carried past vestiges. One could see these trends operating when it came to the body of literature on the physical features and social practices of nomadic caravanners.[24]

[22] Nicholas B. Dirks, *Castes of Mind: Colonialism and the Making of Modern India*, New Delhi: Permanent Black, 2002, p. 44.

[23] Ibbetson argued that tribal groups based on descent had become amalgamated into large tribal groups that were organised around occupations and specific functional affiliation. See Susan Bayly, 'Caste and Race', pp. 165–218.

[24] John Briggs, 'Account of the Origin, History, and Manners of the Race of Men Called Bunjaras', in *Transaction of the Literary Society of Bombay*, vol. I, London: Longman, 1819, pp. 170–197; also see his 'Origin, History and Manners of the Race of Men Called Bunjaras', *The Quarterly Oriental Magazine Review and Register*, I, March–June 1824, pp. 228–31; Edgar Thurston and K. R. Chary, eds, *Castes and Tribes of Southern India*, 7 vols, repr., New Delhi: Asian Educational Service, 1987, IV, pp. 207–236; S. S. ul Hassan, ed., *The*

Nevertheless, what we can grasp from the European travellers' accounts and colonial anthropology concerning communities of nomadic caravanners is that there were several social groups of communities engaged in the caravan trade, and many of them had distinct cultural and historical backgrounds. These texts also suggest that each group, in the beginning, traded in particular foodgrains, as is evident from Tavernier's accounts. In the course of time, they not only expanded their trade activities to other foodgrain items, but also expanded their social base, assimilating many other groups or individuals from various other subaltern castes. It is very much evident in the case of the communities of the nomadic caravanners of south India. They still maintain and assert their groups' early names and culture. For instance, in Hyderabad State they were popularly called lambadas or *lamanis*. It is presumed that the term 'lambada' was derived from the Sanskrit *lavanah* or salt; as they were engaged in salt transportation and trade. Similarly, in Madras Presidency, they were called both lambadas and *sugali*, a term that may be a corruption of the Hindi *supari*, or betel nut, because they were involved in the betel nut trade.[25] The *mukri banjaras* of central India were associated with the maize or corn trade, suggesting that the term *mukri* may be have been derived from the Hindi *makka* or maize.[26] The *mathuri banjaras* of Bombay and the southern part of the United Central Provinces took their name from Mathura as it was presumed that they originated from there.[27] Similarly, all Mahomedan caravanners are said to have come from Multan, and are called *multani banjaras*.[28] A group of caravanners

---

Castes and Tribes of H. E. H. the Nizam's Dominion, Bombay: The Times Press, 1920, pp. 15–27; L. K. A. Iyer and M. V. Nanjundayya, ed., *The Mysore Tribes and Castes*, 2 vols, Mysore: Mysore University, 1928, II, pp. 135–94; R. V. Russel and R. B. Hira Lal, eds, *The Tribes and Castes of the Central Provinces of India*, 4 vols, London: Macmillan, 1916, II, pp. 62–193; Enthoven, *The Tribes and Castes*, II, pp. 331–43.

[25] Thurston, *Castes and Tribes*, IV, 207.

[26] Crooke, *The Tribes and Castes and Castes*, I, p. 160.

[27] Russell and Hira Lal, *The Tribes and Castes*, II, p. 171.

[28] Edward Balfour, ed., *The Cyclopaedia of India and of Eastern and Southern Asia*, 5 vols., repr., Graz: Akademische Druck U Verlagsanstalt, 1967, I, p. 270.

in central India are called *hatadiya* or bullock, as they worshipped the bullock in the past.[29]

The term 'lambadas', which was associated with the caravanners of Hyderabad State, probably first appears in the accounts of Francis Buchanan, who toured south India extensively at the end of the eighteenth century. Buchanan has described them extensively in his myriad accounts while describing their plundering of villages when in the service of the British army.[30] During the same period, John Malcolm, in his reminiscences of central India, calls them *bringarie*s and *lombana*s and uses the terms interchangeably.[31] It is not to suggest that the lambadas' existence as caravanners began from the eighteenth century, but the term appears in the texts of this period. They appear to be identical to the caravanners of fifteenth-century south India.[32] Also, all the 'Banraja carvanners' of south India are identified as one and the same community, although they are called by various names in different regions.[33]

The colonial anthropological notes and European travellers' accounts suggest the mixed social character of the lambada community. As Enthoven says, the genealogy of social and racial compositions of the lambadas was as complex as their homeland.[34] Malcolm supposes that the lambadas were one of the lowest Rajput castes.[35] Most ethnographic notes of the period presume that the origins of a large bunch of south Indian lambadas can be traced to the *charan* and *bhat* castes of western Rajputana and the mathura banjaras from the United Central Provinces. The charan and bhat castes were traditional genealogists during the high noon of the Rajput states. After those states were conquered by

[29] Crooke, *The Tribes and Castes*, I, p. 156.

[30] Francis Buchanan, *Journey from Madras through the Countries of Mysore, Canara and Malabar*, 3 vols, London: Fort Cadell and W. Davies, and Black Parry and Kingsbury, 1807, III, p. 432.

[31] John Malcolm, *A Memoir of Central India, Including Malwa, and Adjoining Provinces, with the History, and Copious Illustrations of the Past and Present Condition of that Country*, 2 vols, repr., New Delhi: Sagar Publication, 1970, II, p. 152.

[32] Briggs, 'Account of the Origin', p. 176.

[33] Iyer and Nanjundayya, *The Mysore Tribes*, II, p. 135.

[34] Enthoven, *The Tribes and Castes*, II, p. 331.

[35] Malcolm, *A Memoir of Central India*, II, p. 152.

the Sultans and Mughals, they lost their patronage and many of them turned to transportation of grain and other forms of livelihood.[36] It is supposed that a large number of charans and bhats first joined the early caravanners as genealogists and protectors of their caravans and eventually merged into the communities of nomadic caravanners as traders and transporters; many of their groups then moved to south India.[37] The Malwa region of western and central India remained the hub of the lambada trade activities throughout the medieval period. It appears that many communities of the regions merged with the lambada caravanners in search of a livelihood. *Kunbi*s (peasant castes), *chamar*s, *mang*s and *mahar*s (agriculture labour and village servants), *nhavi*s (barbers) and downtrodden Muslims of these regions worked with the lambadas as wage labourers. As these communities continually travelled with the lambada caravanners, their ties with their parental castes broke down over a period of time and they eventually merged into the lambada community.[38] Thus the lambada community evolved gradually. Although mythological stories about the lambadas, said to date from the early nineteenth century, claim their origin from Rajput ancestry, those in fact reflect their multi-racial/caste composition.

It is noted in most accounts that the lambadas of Hyderabad State originated from north-west India. The first arrival of the lambadas in the south is mentioned in Ferishta's *A History of the Rise and Progress of the Mahomedan Faith in the Country of Hind*. In 1417, a large convoy of caravanners' bullocks was seized by prince Khan Khanan, the brother of Feroze Shah Bahmuni, when the former rebelled and made an attempt on the throne of Gulburga, the Deccan capital.[39] However, there is ambiguity about their arrival in the south. Crooke suggests that the first mention of banjaras in the south was made in the history of Mohammedans during Sikandar's attack on Dholpur in 1504, and believes that a large convoy of charan banjaras came south with Asaf Jha (also called Asaf Khan), *Wazir* of Shah Jahan, in 1630, under the

[36] M. A. Sherring, ed., *The Tribes and Castes of Rajasthan*, repr., New Delhi: Cosmo Publication, 1975, p. 54; Malcolm, *A Memoir of Central India*, II, pp. 99, 133.

[37] Malcolm, *A Memoir*, p. 170.

[38] Enthoven, *The Tribes and Castes*, II, p. 333.

[39] Ferishta, *History of the Rise*, II, p. 243; also see Briggs, 'Account of the Origin', pp. 174–76.

leadership of Bhangi and Jhangi Naiks of the Rathod clan, with 180,000 bullocks, and that of Bhagwan Das Naik of the Vadtiya clan, with 52,000 bullocks.[40]

In contrast to this, John Briggs presumed that the lambadas had been involved conspicuously in grain transportation in the south from the fifteenth century. They moved into the south largely with Mohammad Tughluk's army, which raided this region several times, as both transporters and soldiers. Among them, large groups came from Ujmeer (Ajmer) under the leadership of Bhecca, who was widely worshipped by the lambadas of Hyderabad. Briggs also notes that their history was not recorded for six generations, merely retaining the names of the lineal descendants from their founder, until the time of Sarung (called Bhangi, from his propensity to imbibe the intoxicating drug called *bhung*) and Bhagwan Vadtiya, the two chiefs under whose leadership the lambadas moved south in large numbers. Sarung was carrying on the transport trade in the region of Dowlutabad, at the head of a large herd of 3,000 bullocks driven by members of his clan. His playfulness resulted in his followers disobeying him and they dispersed in smaller groups all over the Deccan. Vadtiya, in contrast, was thrifty and of steady conduct; he united nearly the whole of his clan and controlled 12,000 head of cattle. Briggs dates this development to the time of the Mughal emperor Aurangzeb's campaign against the rulers of Bijapur and Golconda. Like this predecessors, Aurangzeb employed a large lambada horde in order to supply foodgrains to his army. Vadtiya supplied him with 12,000 men and Sarung Bhangi with 180,000. The latter is said to have mobilised his entire Rathod clan for this purpose. They were also given a patent right to bring grains to the army without any duty and additionally three privileges: to take the thatch from any house when grass was scarce; to appropriate for their own use water which may be drawn and stored in pots ready for a family's use; and to indiscriminately plunder the enemy's country.[41]

Although there is considerable ambiguity in the historical records on the lambadas' trade activities and their arrival in the south, what is clear is that they were traditionally involved in the foodgrain trade and transportation of goods to armies during their campaigns. After

[40] Crooke, *The Tribes and Castes*, I: p. 151.

[41] Briggs, 'Account of the Origin', pp. 177–78.

Aurangzeb's campaign many of them remained in the south, carrying on with their foodgrain trade and transportation. This was because the region offered great scope for the grain trade as it had been visited by recurring famines and drought. Under such circumstances, the transport of grain from one place to another became an occupation of considerable importance. As the Deccan was devoid of a single navigable river[42] and had no roads that allowed wheel-carriages, the entire trade was carried on with the help of laden bullocks.[43] The whole of the northern part of Hyderabad State and many parts of Madras Presidency were also hilly and forested, and even the caravans struggled to cross these regions.[44] There were few good roads in south India even in the first half of the nineteenth century. Enugula Veeraswammayya, who travelled from Chennapatam to Allahabad through Atmakur, Srisailam, Wanparthi, Nirmal, Edulabad and Nagpur towns, describes the woeful condition of the road and transportation system in Hyderabad State during the 1830s. Veeraswammayya says the roads between the Krishna and the Godavari rivers were not suitable even for walking.[45] Also, because of the political situation created by Aurangzeb's death, the region saw intense rivalry and warfare, which proved to be advantageous to the lambadas.[46]

Fragmented Politics, Rivalries and Caravan Trade

As mentioned above, the lambadas of Hyderabad had a long association with commissariat transportation for the army. The same tradition was

[42] The rivers were not suited for long-distance navigation, being either swollen by the seasonal rains, or becoming too shallow in the hot season. Only in short and disconnected stretches were they at all navigable.

[43] Briggs, 'Account of the Origin', p. 172.

[44] *Manual of the Administration of the Madras Presidency, An Illustration of the Records of Government & the Yearly Administration Reports*, ed. C. D. Maclean, repr., New Delhi: Asian Education Service, 1987, I, p. 19; Buchanan has also described how lambada caravans struggled to cross hilly areas in the south. See his *Journey from Madras*, II, p. 18.

[45] Enugula Veeraswammayya, *Kashiyatra Cheritra*, repr., New Delhi: Asian Education Services, 1991, pp. 30–49.

[46] Mark Wilks, *Historical Sketches of the South of India*, repr., Mysore: The Government Branch Press, 1932, p. 518.

continued during the European colonial conquest of India. Nonetheless, during their engagement with the colonial and indigenous rulers, the lambadas lost their relative independence over their trade and became gradually more dependent. In this context, it is important to examine the ways in which the colonial powers treated the lambadas, and how it affected their trade network.

The Mohammedan campaigns in the south started from the late thirteenth century and ended with Aurangzeb's death in 1707 which opened up intensive political rivalry, as the Mughal imperial power began to fade.[47] Succession rivalries and political instability in Delhi led the Governor of Deccan and other local powers to form independent regimes in south India. Once the Governor of the Deccan, Nizam ul Mulk, began to establish his independence, many other rulers also began asserting themselves, either by acquiring grants from the Delhi ruler or by declaring themselves independent rulers. The Peshwas established themselves in the Maratha region, and the Nawabs of Mysore in the Karnataka region under Haider Ali and Tippu Sultan. The Hindu Rajas of Tanjore and Travancore asserted their power in the far south. These developments created political rivalries and tensions throughout the region, as these newly emerged states vied with each other, and with the emerging European powers, to extend their territories and influence.[48]

It is needless here to go into the details of this politics, but by the middle of the eighteenth century, the Peshwas of Poona, the Nawabs of Mysore and the Nizams of Hyderabad stood out as the main rival powers in the south. Each was waiting for an opportunity to wage war against the other to extend its sovereignty in the region. The death in 1732 of the Nawab of Karnataka (the region was also known as Arcot), who was a subordinate of the Nizam, resulted in a war of succession. A conflict was also witnessed when the first Nizam died in 1748. These wars led to intense rivalries among the rulers of the south, as each party was supported by the states of the region and some Europeans. Anglo–French rivalries in Europe, which were extended

[47] J. D. E. Gribble has given a detailed account of Mohammedan invasions of south India. See his *History of the Deccan*, 2 vols repr., New Delhi: Mittal Publications, 1990, I.

[48] William Lee Warner, *The Protected Princes of India*, London: Macmillan, 1894, pp. 63–80; also see Sarojini Regani, *Nizam–British Relations, 1724–1857*, Secunderabad: Swarajya Printing Work, 1963, Preface, p. ii.

to India, added another dimension to the political situation. The three Karnataka wars between 1745 and 1763 saw the British emerging as the foremost power in the south. The Nizam concluded a treaty with the British in 1766 and joined with them against the rising power of Haider Ali and Tippu Sultan. This resulted in four Anglo–Mysore wars between 1766 and 1799, which ended with Tippu's death in battle and the annexation of a major part of Mysore by the British. The Nizams were also in alliance with the British in three Anglo–Maratha wars in the years 1775–82, 1803–05 and 1817–18. After these wars, almost the whole of the Deccan came under the direct control of the British. The only major territories not controlled by them directly after 1818 were the states of Hyderabad and Mysore.[49]

In all these wars, the lambadas were the main grain suppliers to all sides. Their services were acknowledged by all the imperial powers. They had particularly good relations with the Nizams, as they had served their armies from the times of Aurangzeb. The first Nizam employed them in large numbers during his early consolidation of the Deccan. He issued a special order stating:

Ranjan ka pani,
Chhappar ka ghas,
Din ka tin khun mu'af;
Aur jahan Asaf Jan ke ghore,
Wahan Bhangi Jhangi ka bail.[50]

This means 'if you can find no water elsewhere, you may even take it from the pots of my followers; grass you may take from the roofs of their huts; and if you commit three murders a day I will even pardon this, provided that wherever my cavalry is stationed, there I always find Bhangi Jhangi's bullocks'. Almost all of the ethnographic notes of the nineteenth century state that this order, which was engraved in golden letters on a copper plate, was in the possession

[49] Hastings Fraser, *Our Faithful Ally, the Nizam: Being an Historical Sketch of Events, Showing the Value of the Nizam's Alliance to the British Government in India and His Services during the Mutinies*, London: Smith, Elder and Co., 1865, pp. 27–97.

[50] Syed Hussain Bilgrami and C. Willmott, *Historical and Descriptive Sketch of His Highness the Nizam's Dominions*, Bombay: The Times of India Steam Press, 1883, p. 344.

of the lambadas of Hyderabad. It is also noted that the British had issued similar orders.[51] The first Nizam constructed a separate gate for the lambadas' herds to pass into the Golconda Fort.[52] There was also the practice of presenting the descendants of Bhangi (chief) with a *khillat* (a gift of turban cloth) from the Nizams of Hyderabad. It was considered a mark of great honour, when the Nizam used to present eight *thans* of *khadi* (a *than* equal to 16 yards) every year, which the chief *naik* (headman or chief) used to tie his turban.[53] The British also continued this practice; after every war, the lambada naiks and leaders were honoured with clothes, turbans and swords.[54] Whatever the façade, it was a kind of taming through which the imperial powers ensured the subordination of the lambadas.

In the Karnataka wars, the lambadas were first employed by the second Nizam, Nasir Jung, who gave a patent right dated 15 *Zeehuj* 1162H (15 November 1749) to Narayan Bhangi to furnish his army with supplies. Later, during the war, they were also employed by the enemy, the French, in the same capacity. In fact, when Bussy, the French chief commander, was losing ground in the Karnataka war, he used the lambada horde residing at Char-Mahal in Hyderabad to protect him from the Nizam's army from 5 July till 15 August 1755. The first time that the lambadas were employed by the British was during the war with Tippu Sultan in 1791–92, when the British commissariat arrangement failed. During that period, the allies, consisting of the Hyderabad and Poona armies, co-operated with the British troops from Madras, under the personal command of the Governor General of India, Cornwallis. In such a situation, it was the timely supply of provisions by the lambadas that saved the British army from defeat. Bheema Naik Bhangi received a cash advance and a patent from Sir John Kennaway, the British Ambassador at Hyderabad; and his brother Somah received a similar reward from Cornwallis himself.[55] Colonel

[51] Crooke, *The Tribes and Castes*, I, p. 151.

[52] It is noted that the banjara gate of the Golconda Fort had a special attraction in many ways among all the eight gates. See Bilgrami and Wilmott, *Historical and Descriptive Sketch*, p. 511.

[53] Hassan, *The Caste and Tribes*, 21.

[54] *The Mysore Letters and Dispatches of the Duke of Wellington, 1799–1805*, Bangalore: Mysore Government Press, 1862, p. 68.

[55] Briggs, 'Account of the Origin', pp. 182–83.

Mark Wilks gives an interesting description of the deployment of the lambadas by Cornwallis. He recounts that about 60,000 lambada oxen had furnished the British army with grain from various parts, but chiefly from Coromandel.[56] Arthur Wellesley too employed the lambadas massively in the last Mysore War and the second Maratha War. In fact, he was always engaged in collecting lambada suppliers.[57] The Peshwas of Poona and Satara also employed them throughout their campaigns against the British and the Nizam.[58]

The usual practice when engaging the lambadas for commissariat transportation was to give them an unsecured advance to purchase grain from any part of the country. A patent order was simultaneously issued stating that the said party was recruited by the army and should not be stopped from procuring grains in any territory and that no duty should be levied on the grain. Upon their return, they were bound to give an account of the quantity of grain purchased, and were not permitted to open a single bag until they had obtained the permission of the commander-in-chief. They had to sell their goods at the average rate of the camp-market. After the disposal of the entire stock, they had to repay the original sum advanced to them and could only then distribute the remaining amount among themselves for their labour.[59] Such an arrangement sought to satisfy the mutual interests of both parties. According to Briggs:

> I have before premised that the English commander advanced large sums of money on this occasion to the Bunjaras (Lambadas), and I have reason to suppose that it was a thing which had never before been done, but what security he received for this advance, or for the due performance of the contract, I am unacquainted with; it was probably a compact of honour, but so blended with mutual interests that the parties had confidence in each other.[60]

A similar view is also presented in recent studies of the eighteenth century. Referring to the political rivalries of this period in south India,

[56] Wilks, *Historical Sketches*, p. 517.
[57] *The Mysore Letters*, p. 84.
[58] Bilgrami and Wilmott, *Historical and Descriptive Sketch*, p. 338.
[59] Briggs, 'Account of the Origin', pp. 186–87.
[60] *Ibid.*, p. 184.

P. J. Marshall has stated that 'these wars were notoriously destructive, but the movements of huge armies created many opportunities for the enrichment of the Indians who supplied them, especially the Banjaras who followed the armies with their stocks of grain and other supplies.'[61] Recent studies of eighteenth-century India have emphasised the positive side to the break-up of the Mughal Empire and the emergence of regional successor states, and it appears that the lambadas and banjaras were in a favourable position to benefit from this process. The wars in particular provided opportunities to the lambadas and others because they created a number of army camp markets, Army commanders extended protection to merchant communities from enemy forces during the war.[62] The situation was not, however, entirely advantageous to them.

To start with, the lambadas were relatively independent merchants and transporters and had their own trade practices. Peter Mundy, a European traveller of the seventeenth century, stated that they (lambadas) were basically independent traders buying foodgrains from rich areas and selling them in scarcity areas.[63] But once they entered into a contract with the British colonisers, their trade activities were strictly regulated and monitored by army commanders and they lost their independence and monopoly over their trade. They were not allowed to sell the procured grain to either enemy forces or civilians. From Arthur Wellesley's correspondence it is reported that 'the Binjaries drop in by degrees, and I get a little from the country; but I have determined that not more Binjary rice shall go into the bazaar.'[64] In another instance he says that 'the Banjaras I look upon in the light of servants of the public, of whose grain I have a right to regulate the sale.'[65] Although pre-colonial rulers had also regulated and monitored their trade, the patents issued to them by the Mughals and the Nizams suggest that they were given a freer hand in many ways previously, as they were allowed to determine the source from which to procure grain, to ensure reasonable prices and given a right to obtain fodder for their cattle.

[61] P. J. Marshall, ed., *The Eighteenth Century in Indian History: Evolution and Revolution*, New Delhi: Oxford University Press, 2003, p. 1.

[62] Siddiqi, *Trade and Finance*, pp. 40–41.

[63] Richard Carnac Temple, ed., *The Travels of Peter Mundy in Europe and Asia, 1608–67*, 5 vols, London: Cambridge University Press, 1914, II, p. 95.

[64] *The Mysore Letters*, p. 141.

[65] Russell, *The Tribes and Castes*, II, p. 170.

Market prices were generally higher than military camp prices, and this used to tempt the lambadas who were carrying grain and other goods to the camps. The British set out a number of regulations to check such sales, and held a breach of them to be a serious crime. From the time of the last Mysore war, a native commissary of the East India Company always accompanied the caravans to ensure that there was no such breach of contract.[66] During this war, the allied forces employed huge caravans. The British were served by 35,000 pack-bullocks under the leadership of Bheema Naik Bhangi, the Nizam by 26,000 pack-bullocks under the leadership of Latchu Naik Bhangi, and various commanders a further 23,000 under other leaders.[67] A special army unit under an *amildar* (subordinate military officer) closely monitored each caravan.[68]

Fluctuations in market prices during wars had a serious effect on the lambadas' caravan trade and they could depress their profits, as prices often went up at the start of a war and fell at its end. As a result, the lambadas often incurred losses. This was the case during the last Mysore war. When this war was about to break out, the services of lambadas were employed. The British Resident at Hyderabad advanced Rs 150,000 to their chief, Bheema Naik, at Hyderabad to supply grain to the army marching towards Seringapatam. Before capturing Seringapatam, there were distress conditions across the region and the grain prices had suddenly shot up. One *seer* (one kilogram) of rice was being sold at Rs 2. Even the army experienced a serious food shortage. The Nizam's army under the command of Colonel Sir John Malcolm had 25,000 bullock-loads of grain, but it was stuck below the ghats. The army under the command of General Harris experienced considerable distress during its advance to Seringapatam for want of grain. The general heard that Bheema Naik with a supply of 15,000 bullock-loads was at the foot of the ghats and sent a detachment to protect him from the enemy forces. But by the time Bheema Naik's bullocks reached the destination, the British army had captured Seringapatam. They opened the large granaries of Tippu and that brought down the price to 30 seers of rice for a rupee. However, the British general ultimately adhered

[66] NAI, HRR (misc.), vol. 225, 1 January 1800–2 December 1801, p. 357.
[67] John Briggs, 'Account of the Origin', p. 187.
[68] NAI, HRR (misc.), vol. 225, 1 January 1800–2 December 1801, p. 357.

to the contract and purchased the entire stock at five seers for a rupee. But if the Naik had reached before the fall of Seringapatam, he would have made a greater profit. It was also possible that had he reached even later he would have been ruined, as the rates were falling rapidly.[69] For the lambadas this was a fairly regular experience when supplying to the army. The army records show that the lambadas had to occasionally demand payment at appropriate rates, as the army commander who was supposed to make the payment was always busy with war.[70]

The British forces resented the lambadas' monopoly in foodgrain transportation and treated them harshly, especially after the last Mysore war. The Chief Commanding Officer of the force in 1801 submitted a strong proposal to the Chief Secretary of Government of India to bring changes in the agreement with the lambada carriage. As per the new agreement, it was decided to pay them at the wholesale market price, which was Rs 3 for one bullock load.[71] It obviously cut down their profits, as the wholesale price was less than the retail one. The lambadas usually purchased the grains from the villages and were often ignorant of prices in the urban centres. This sometimes caused them great losses. The commanding chiefs and other officials also proposed that they establish their own carriage system to break the lambadas' monopoly.[72] Arthur Wellesley was particularly suspicious of the lambadas, as is seen from his field notes. He suggested to the commanding officer not to make their purchases duty-free, presuming that they would take advantage of this concession in the future or after the war period.[73] The army took it for granted that the bullocks would be accessible at all times during a war. It was very difficult for the lambadas to feed their cattle, as fodder was expensive during the war. Indeed, most of their earnings were spent on feeding the cattle, which left very little for their subsistence, as the commanding officer used to force them to accompany the army.[74]

[69] Briggs, 'Account of the Origin', p. 186.

[70] NAI, HRR (misc.), vol. 3, 2 December 1791–24 June 1794, p. 9.

[71] NAI, HRR (misc.), vol. 225, 1 January 1800–2 December 1801, p. 359.

[72] *Ibid.*, p. 365.

[73] *The Mysore Letters*, p. 77.

[74] *The Dispatches of Field Marshal the Duke of Wellington KG, During his Various Campaigns in India, Denmark, Portugal, Spain, the Low Countries, and France from 1799 to 1818*, compiled from official and authentic documents by Lt. Col. Gurwood, 6 vols., London: John Murray, II: pp. 308–09.

The lambadas were also accused of looting war-stricken regions. Depredation of the enemy's country at the end of the war was a regular practice by armies at this time, and the lambadas, too, were certainly culpable of this.[75] After the capture of Seringapatam, a small horde of lambadas in the employment of the British army attempted to leave, and suspecting that they intended to loot the conquered region, the Subsidiary Forces intercepted them and hanged seven of their principal naiks.[76] The armies of the British and the Nizams carried out many such brutal hangings, especially after the last Maratha War of 1817–18.[77] The Peshwas, too, as mentioned above, employed about 40,000 lambadas in their campaign against the British for foodgrain supply. The lambadas had particularly good relations with the Marathas as the latter gave them a free hand in procuring grain and paid them high prices.[78] After the defeat of the Peshwas, many local zamindars and talukdars formed armed bands with those lambadas and fought against the British expansion into their territories. The British saw this as freebooting, to be crushed with a ruthless use of force. Many lambada naiks involved in such resistance were hanged publicly.[79] This led to them being stigmatised by the British as a predatory community—a stigma that haunted them throughout the colonial period.

After the Maratha wars, the British created its own commissariat force that hired its pack animals and supplied foodgrains to their army. They also continued to hire many lambadas as transporters.[80] These arrangements came to an end after the railway connected the major army stations later in the nineteenth century. Similarly, the lambadas also lost their traditional commissariat transportation rights with the Nizam. The Company formed the Hyderabad Contingency in 1812, consisting of British soldiers and the Nizam's army, to assist the Nizam. After the last Maratha war it was encamped permanently at

[75] Buchanan narrates innumerable incidents of lambada depredations in Karnataka. See his *Journey from Madras*, II, pp. 143–44, 150, 165, 175, 186, 190; III, pp. 432, 435.

[76] Briggs, 'Account of the Origin', p. 186.

[77] Reginald George Burton, *History of the Hyderabad Contingent*, Calcutta: The Superintendent of Government Printing, 1905, p. 75.

[78] *The Mysore Letters*, p. 140.

[79] NAI, HRR, vol. 568, part II: 1782–1888, p. 378.

[80] NAI, HRR, vol. 64, 1782–1888, p. 279.

Bolaram (Secunderabad), Aurangabad, Hingoli, Ellichpur, Mominabad, Gulbarga, Warangal, Mucktal, Lingasugur and Jalna and hardly moved, except during the 1857 revolt.[81]

With the loss of their chief livelihood, the lambadas became financially vulnerable and in time fell into the clutches of big traders and merchant moneylenders. Malcolm, writing in the 1820s, observed that the *Hoonda Barawallahs*, who were both moneylenders and contract carriers to big merchants, were giving lambadas advances at high interest rates. Once under their control, they could force the lambadas to provide transport for them at very low rates. Malcolm also noted how the lambada caravans suffered losses from being harassed at the customs houses at each territorial border.[82] Although this had been a widespread phenomenon in the caravan trade also in the pre-colonial period, the lambadas had previously enjoyed enough economic strength to weather such demands. In all, their reduced circumstances under colonial rule meant that they now lacked the resilience to defend their interests effectively.

Organisation of the 'Thanda' and Caravan Trade

It is important here to discuss the formation of the *thanda* (lambada camp/settlement), which was a crucial element in the operation of the caravan trade. The lambada community was divided into thandas based on their transportation and trade engagements, and each thanda had a monopoly over a particular item and region. The thanda was also crucial in the making of the lambada community as it constituted a common social and moral order of the caravan people. A thanda was a trading group and its naik (chief) was the agent for the group's trade and transportation engagements. The literal meaning of 'thanda' is 'a camp', comprising a large number of people with pack bullocks.[83]

[81] Henry George Briggs, *The Nizam: His History and Relations with the British Government*, 2 vols. repr., New Delhi: Manas Publication, 1985, II, p. 162.

[82] Malcolm, *A Memoir of Central India*, I, p. 95.

[83] Hassan, *The Castes and Tribes*, p. 15.

It finds a mention in the writings of many late medieval travellers and colonial officials. Peter Mundy writes that he met a huge thanda caravan of grain transporters on 14 December 1630 on his way to the Deccan from Burhampur.[84] Colonial Tod also used the term 'thanda' and 'Banjara' interchangeably.[85]

It was, as mentioned above, the general practice in trade agreements for one member of a thanda, generally the chief naik, to take the initiative and draw up a contract with a king or his subordinate ruler. According to the agreement, the chief naik would mobilise and employ other lambadas from different groups. If required, he would also employ non-lambadas from various lower castes that engaged in manual labour in villages. Each thanda was again divided into small groups and put under a sub-leader (sub-naik). The sub-naik's position was generally given to the individual who mobilised or gathered that particular group. He was instrumental in carrying on trade. The entire thanda was the proprietor of the general stock that they procured for trade or transportation purposes, and they shared the profits after settling the advance that they received from their employers. However, the chief naik was considered to be the chief proprietor of the caravan and had the exclusive right of appropriating to his own use everything that was presented to him by his employers. He was seated on a *gadi* (seat of power) created by Sarunga Bhangi who came to the south with Aurangzeb's army. There were only four chief naiks at the end of the eighteenth century. Each main clan had one chief naik. Such a 'clan' did not necessarily imply kinship or a blood relationship, but rather represented a group of people gathered under a leader for the purpose of grain transportation. The main clans were Rathod, Vadtiya, Power and Chowhan. Although the chief naiks were supposed to gain their authority from the assent of their followers, in time the post had become hereditary, being generally passed on to the eldest son. In practice, they held absolute power over their people. A *karbhari* (secretary), who was elected by the clan as a whole, assisted them.[86]

Although the transporting of grain was largely monopolised by the lambadas, and operated with the community's set of rules, there were

[84] *The Travels of Peter Mundy*, II, p. 54.

[85] James Todd, *Annals and Antiquities of Rajasthan*, ed. William Crooke, 3 vols, London: Oxford University Press, 1920, III, p. 751.

[86] John Briggs, 'Account of the Origin', pp. 178–79.

frequent fights between the chief naiks for patent or commissariat rights for army transportation. Indeed, their employers often stirred up rivalries between the clans for their own advantage. When in 1749 Narayan Bhangi Rathod was given a patent by the Nizam, Vadtiya Naik demanded a similar patent, to no avail. This fanned the hatred between the two groups and violence ensued. When a large group of Rathods camped in the neighbourhood of Badrachalam, situated in the vicinity of the Godavari river, lying to the north-east of Hyderabad, the Vadtiyas attacked them. The Rathods suffered severely, but managed to take one female Vadtiya captive. A few months later they came into conflict again at a place called Ramagoorum, on the banks of the Krishna river. A violent fight ensued in which the Rathods were completely defeated, suffering severe losses in terms of both men and property; above all they lost the order of patent right given to the Rathods by Aurungzeb. This is said to have happened in the time of Bussy, in the middle of the eighteenth century. Although it was the last dispute of its kind, the practice of giving presents to the thanda naiks by the British and the Nizam continued to create rivalries between the groups.[87]

The lambada caravanners were divided into four clans or thandas, but they followed common social codes. In the course of time, the thanda became a distinguishing feature of the lambada community as they evolved their own cultural and belief systems and language. The thanda constituted and reflected every element of their lives: power, politics, clan and lineage, domination and subordination, culture, religion and identity.[88] Thus the thanda was a multi-layered and multipurpose system that developed in the course of their nomadic life and continued in their settled lives. A set of social codes was probably drawn up at the beginning of the eighteenth century, during the period of Sarunga Bhangi, to govern all the thandas.[89] The clan was the principal social base of a thanda, even when the lambadas scattered into small groups and settled permanently in the later period. Each thanda consisted of only one clan. Those who joined the thanda in search of a livelihood

[87] A. C. Lyall, *Gazetteers for the Haidarabad Assigned District, Commonly Called Berar* (Bombay: Education Society Press) 1870, p. 196; also see Briggs, 'Account of the Origin', p. 182.

[88] Malcolm, *A Memoir of Central India*, II, p. 152.

[89] Briggs, 'Account of the Origin', p. 179.

from other castes had to take on the clan name of the thanda. The practice was to assimilate 'outcastes' into the thanda community. There is insufficient proof to suggest when this custom actually evolved, but it is widely mentioned in most of the ethnographic notes of the late nineteenth century. According to this custom, after a few years the new member, called the *jangad* (outcaste absorbed into the thanda), would give a feast to all the members of the thandas. The members of the thanda would partake of the meal and declare the newly enlisted person to be a member of the thanda; and he was given the thanda's clan name. After this ceremony the jangad was treated as an equal and allowed to enter into matrimonial relations with the lambadas. Thus, many people from other communities joined the lambada community. The thanda also had a group of entertainers called the *dhadi*, *bhat* and *dapadya*, who were an integral part of the thanda and would amuse the thanda people on all social occasions. The jangads and the entertainers occupied the lowest rung in the thanda hierarchy.[90]

The naik's power over the thanda being total, nothing could take place there without his knowledge. One had to take his permission even to have marital relations with someone from another thanda. He was the sole executive and judiciary of the thanda. His role in settling criminal and civil disputes was crucial. The cases that were placed before the thanda panchayat generally dealt with marital and sexual concerns. In the early caravan days, promiscuous sexual relations were common and these often resulted in violence. The clan barriers were not observed when it came to marital relations, and sometimes they also married outside their own lambada community. To avoid confusion, Sarunga Bhangi established a code that if a clan member was guilty of marrying within the clan, the act should be considered as incest, and the criminal expelled from the thanda and the community. From then onwards, a clan with the same name was considered a fraternity, and women, though not the men, were expected to be chaste.[91] The same institution continues in the lambada community today. According to the codes of the thanda panchayat, a person could not be punished in any way without being first tried by a jury of five, consisting of the naik and others of the thanda, who were selected by the accused

[90] Russell, *The Tribes and Castes*, II, p. 183.

[91] Briggs, 'Account of the Origin', p. 179.

or both the accused and the victim. The jury held long discussions before giving its judgment, and the agreement was considered binding on both parties. Despite this, swords were sometimes drawn, and the onlookers had to intervene to stop a fight. It was the practice that the man who thought the issue settled took off his turban, and threw it full length between the parties. The parties would then have to agree to his suggestion.[92]

The harshest punishment was expulsion from the thanda. The guilty party was taken to the outskirts of the thanda, beaten with his own *cheppal* (footwear) four times and expelled. None in the same thanda or no group from an established thanda was to receive him; those who violated the rule were called by the panchayat and fined. In most cases, the guilty were punished by fines: petty crimes were fined Rs 5, the intermediate fine was Rs 12, and the most severe fine was Rs 30. The entire amount was generally spent on bhung and *daru* (country liquor). The thanda panchayat would not usually wish to impose a death sentence, except in the case of murder. The sentence was executed by friends and relatives of the victim, who had to put the murderer to death within a period of three days after the deed was committed, after which no one should try to take revenge.[93]

A writer in the East India Company at Bombay, James Forbes, described a lambada thanda in the last quarter of the eighteenth century. The thanda consisted of a set of merchants who did not belong to any particular territory. They lived in tents unitedly for mutual comfort and safety in the transportation of their merchandise. They were governed by their own laws.[94] Thus the values practised among the groups of thandas not only served to organise and carry out the caravan trade efficiently, but also created a community of caravanners.

When the thanda embarked on procuring grain, the entire group would move together, including the families. Before entering a war zone, however, they often lodged their families at suitable distant places.[95] The length of a thanda when travelling was about one-and-

[92] *Ibid.*, p. 180.

[93] *Ibid.*, p. 181.

[94] James Forbes, *Oriental Memoirs: A Narrative of Seventeen Years Residence in India*, 2 vols. repr., London: Richard Bentley, 1834, II, p. 276.

[95] Wilks, *Historical Sketches*, p. 517.

a-half miles. They travelled six to seven miles a day.[96] This train often travelled with loaded bullocks for about 1,500 to 2,000 miles during times of war or when there were fairs to sell and procure foodgrains.[97]

The lambadas' caravan trade was crucial in building a market network in rural Hyderabad. Although large groups of lambadas were involved in supplying foodgrains to the army, many of them traded independently. After they lost their traditional commissariat rights with the army in the early nineteenth century, many lived by trading in foodgrains in the rural areas of Hyderabad State and the adjacent British territories.[98] At the beginning of the second decade of the nineteenth century Briggs noted that the four clans of the lambadas were well spread out. The Rathod clan possessed 90,000 pack-bullocks spreading from the head of the river Wurdh in Gondwana to Cuddappah through Nirmal, Hyderabad, Kurnool, and the confines of Mysore. They were distributed as follows: 50,000 in Hyderabad, 4,000 in Aurangabad, 24,000 in Poona and Konkan and 12,000 in the ceded districts. The Vadtiyas possessed 80,000 pack-bullocks and were spread between Chicacolem and Nellore of the Madras Presidency. The Chowhans possessed 18,000 pack-bullocks, occupying a line of country between south of Nellore and modern Karnataka. The Powars were the least significant, possessing only 12,000 pack-bullocks, trading mostly between Ganjam and Gondwana. Although associated with particular regions, each group was frequently on the move, ranging all over south India.[99]

The lambadas largely traded in foodgrains and salt, employing a sort of barter system. They carried salt, rice, piece goods of native manufacturers, raw silk, spices, oil-seeds, sugar, dried coconuts, jaggery, tobacco, salt-fish, opium, and betel leaves from various ports of coastal Madras Presidency to the interior part of Hyderabad State, and, in return, took cotton, foodgrains like wheat, *ragi*, *jawar* (millets) and other millets, pulses, tamarind, wax and iron.

[96] *The Travels of Peter Mundy*, II, p. 36, 95.

[97] Forbes, *Oriental Memoirs*, II, p. 277.

[98] Maulavi Cheragh Ali, *Hyderabad (Deccan) Under Sir Salar Jung*, 4 vols., Bombay: Education Society Press, 1886, IV, p. 397; Forbes, *Oriental Memoirs*, I, pp. 276–77; also see Vaikuntham, *State, Economy and Social Transformation*, p. 159.

[99] Briggs, 'Account of the Origin', pp. 192–93.

The lambadas used to sell their procured goods and grain at advantageous rates in the rural areas on their way between two territories. When their herds passed through the villages, the people bought salt, foodgrain and other items from them. If anything remained after their long journey, they would sell them at the weekly fairs in the Hyderabad State and coastal districts of Madras Presidency.[100] The *Manual of the District of Vizagapatam* noted that 'there is a market held in almost every village once a week at which you find the Brinjaries and low country traders making their bargains.'[101] Similarly, weekly fairs were the main network of rural marketing in Hyderabad State. There were about 250 weekly fairs in Telangana and about 700 in the Marathwada regions. There were also about 2,437 *uruses*, *mela*s and *jatara*s (fairs and pilgrimages) in the state where there was a large volume of trading. The lambadas used all of these markets to exchange the goods that they brought from the coastal regions.[102]

The lambadas also acted as hired transporters to the East India Company and the native merchants. In fact, they carried on the whole of the external trade between Hyderabad and Madras. Naryanapuram, between Hyderabad and Masulipatam, which was the main trading centre of the Company from the late eighteenth century, was a great emporium of trade and had several warehouses and godowns for storing goods in transit. The lambadas' bullocks were employed by the traders to transport finished silk, *chintze*s, tin, copper, carpets, rose water, fine cloth and other products of Masulipatam, Bengal, Ganjam, Vizagapatam and Chicacolem from Naryanapuram to the Nizam's country and, in return, they brought articles like wax, honey, cotton and wheat.[103] The hiring rates offered by the Company and merchants varied from place to place. In Vizagapatam and Vizianagaram, the lambadas were offered four *anna*s (an anna was sixteenth part of a rupee) for a bullock for nine to 10 *coss* (miles) at the beginning of the nineteenth century.[104]

[100] Balfour, *On the Ethnology of Hyderabad*, pp. 10–12; also see A. V. Ramana Rao, *Economic History of Andhra Pradesh, 1766–1957*, Bombay: Popular Book Depot, 1958, p. 130.

[101] *A Manual of the District of Vizagapatam in the Presidency of Madras*, repr., Hyderabad: District Gazetteers Andhra Pradesh, 1994, p. 121.

[102] Vaikuntham, *State, Economy and Social Transformation*, p. 159.

[103] Ramana Rao, *Economic History*, p. 44.

[104] *Ibid*., p. 49.

Regulated Market and Marginalisation of the Lambada Caravan

The question here is how such a well-organised trade was destroyed, with its practitioners being reduced to acute poverty. Most ethnographic studies claim that the marginalisation of the lambadas was brought about by their ingrained cultural values. Their wandering habits, religious practices and habitual drunkenness were seen to be at the root of their poverty and ruin.[105] This was a typical Orientalist argument, and it is not given much credence by more recent historians who argue that the caravan trade was destroyed primarily by the emergence of modern forms of transport, particularly the railway.[106] C. A. Bayly's study on north India suggests that the nomadic groups, including the banjaras, lost their livelihood with the decline of the Mughal Empire, and they settled as either agriculturists or pastoral cattle breeders in the eighteenth century and played an important role in agricultural production.[107] D. A. Washbrook's study on the south observed that the construction of roads for military purposes undermined the lambadas' economy. After they lost their transporting occupation, the lambadas settled down as agriculturists in the forested areas and greatly contributed to the expansion of agriculture in the eighteenth century.[108] It must be mentioned that these eighteenth-century studies largely portrayed the positive side of the historical developments, and they fail to bring out the magnitude of the lambadas' loss which was not brought about only by the decline in the caravan trade. Indeed, they continued to practise this trade in a reduced way right down to the last quarter of the nineteenth century. My argument here is that although the introduction of railways hastened the destruction of the caravan trade, the grounds for its destruction were laid by changes in the market economy that were brought about

[105] Briggs, 'Account of the Origin', p. 193.

[106] Varady has argued that the two major developments that contributed to the eclipse of banjaras as large-scale transporters were metalled roads and the railway system. See his 'North Indian Banjaras', p. 5.

[107] Bayly, *Rulers, Townsmen*, pp. 219–26.

[108] David Washbrook, 'Progress and Problems: South Asian Economic and Social History 1720–1860', *Modern Asian Studies* 22: 1, 1988, p. 80.

by colonial taxes and regulations from the early years of the nineteenth century onwards.

For example, new salt regulations and taxes imposed by the East India Company hit the lambadas hard, as the trade in salt was central to their whole trading cycle. In the past, they had carried salt from the coastal regions that stretched from Ganjam to Nellore, where local castes produced it in extensive saltpans. They exchanged it for cotton grown in the Marathwada region, which was emerging as one of the foremost cotton-growing areas in India.[109] The cotton industries in the Andhra region largely depended on the cotton produced there, and much was also exported beyond the region and even abroad.[110] This trade had provided the lambadas with a very profitable livelihood.[111] The salt trade was brought under the Company's monopoly from 1215F (1805–06). Salt farms and factories were established along the coast that were supervised by special officers or sub-collectors and strictly regulated.[112] Subsequently, salt prices increased abruptly. In 1802, the price had been between Rs 28 and Rs 35 per *garce* (120 Indian maunds). The price increased to Rs 70 per garce in 1805, Rs 105 in 1809, and Rs 180 in 1844.[113] In addition, heavy custom duties were imposed and the caravans were frequently stopped in transit. The lower officials at the saltpans fraudulently weighed salt while selling to buyers, taking some *mamul* (bribe) from the buyers. This practice was well known and the custom officials began stopping the lambadas at every *chowki* (customs outpost) to detect the fraud or excess weighing. Paying mamul at every custom chowki was a regular practice, sometimes the custom official would plunder the lambada caravans if they did not offer mamul. These practices interrupted the lambadas' trade. The extortions doubled the cost of salt by the time the caravan reached Hyderabad.[114]

[109] Maulavi Cheragh Ali, *Hyderabad (Deccan)*, IV, p. 397; also see Rao, *Economic History*, p. 144.

[110] Fraser, *Our Faithful Ally*, pp. 351–72; also see Vaikuntham, *State, Economy and Social Transformation*, p. 161.

[111] *Ibid.*, p. 41.

[112] *Gazetteer of the Nellore District*, Madras: Government Press, 1942, p. 289.

[113] *A Manual of the District of Vizagapatam*, p. 291.

[114] NAI, HRR, Political, file no. 3/1860–65, 1; also see Rao, *Economic History*, p. 135.

Many lambadas stopped going to the saltpans from the 1820s. This had a serious impact on the weaving industries of the coastal districts as they were largely dependent on the lambadas' cotton supply. Realising the situation, the Madras government reduced the salt price from Rs 105 to Rs 70 per garce in 1820. But this did not attract the lambadas as the customs officials still extracted heavy bribes.[115] The sale of salt in the farms also fell drastically. From 1800 to 1822, salt sold to the lambadas at the Masulipatam saltpans was about 5,000 maunds annually, which was a drastic decrease compared to previous years.[116] In Nellore and Ongole districts, the salt revenue decreased from Rs 534,779 to Rs 249,714 between 1805 and 1807.[117] Apart from the customs officials' extraction, there were a number of other restrictions on the lambadas' movement. As there were no proper roads in the coastal districts, the lambadas' caravans had to pass through the fields, and when they crossed certain districts with their loaded cattle, they sometimes destroyed the grain that was growing there. Hence, in places like Masulipatam, they were allowed to travel only through certain villages during certain specified periods of the year, ranging from 24 September to 6 June.[118] Thus the monopolistic high prices, customs exactions, measurements malpractices and other restrictions began to hamper the lambadas' caravan within the British territories.[119]

The lambada caravans also encountered a similar hostile reception when they entered Hyderabad State. Before we discuss the ways in which the lambadas' trade activities were undermined, it is important to look at the deployment of the Commercial Treaty of 1802 between the Nizam and the British Indian government. It was under this treaty that the state's market regulation and economy were restructured along modern lines. It played an important role in destroying many long-standing market relations as well as promoting the emergence of new trading communities in the state. Above all, the treaty made the Hyderabad economy perpetually dependent. The British undermined the trade and prosperity of many Princely

[115] *A Manual of the District of Vizagapatam*, p. 290.
[116] Rao, *Economic History*, pp. 144–45.
[117] *Gazetteer of the Nellore District*, p. 291.
[118] Rao, *Economic History*, p. 39.
[119] *Ibid.*, p. 137.

States through customs policies that benefited trade in British areas at the expense of those states.[120]

The Hyderabad Residency Records of the eighteenth and early nineteenth centuries reveal that market regulations hardly existed in the state at that time. The British, however, considered the existing market practices to be irrational and an obstacle to healthy trade and commerce. In particular, they accused the *jagirdars* and *maktadars* (traditional landlords) of hampering trade activities in the state.[121] Bearing this in mind, they forced the Nizam to implement new market regulations. The preamble of the 1802 treaty stated that 'a well regulated commerce is essential to the opulence and prosperity of the people and to the wealth and power of the state and that a free and secure commercial intercourse tends to maintain and improve the relation of amity, peace and concord between contiguous nations.'[122] Whether such regulations added to the prosperity of the people and the region is a moot point, but they definitely helped the colonial power to squeeze a surplus from the productive resources of the state. The Residents at Hyderabad played a key role in moulding the Nizam's and his prime ministers' views about the need to establish new market regulations and make trade more open and profitable.[123] A number of regulations were laid down in the treaty to regulate market activities between the two territories. According to the third article of the treaty, transit duty was abolished on all articles transiting from one part of the British territories to another through Hyderabad State and vice versa. According to the fourth article, *rahdari* (toll) duties and local imposition were abolished. Article 5 of the treaty restricted the imposition of duty to 5 per cent on

[120] David Hardiman shows how long-standing economy and market relations in the princely states of Rajasthan and Saurashtra were undermined by the colonial state in the early nineteenth century in order to benefit trade in British areas at the cost of the princely states. See his *Feeding the Baniya: Peasants and Usurers in Western India*, New Delhi: Oxford University Press, 1996, pp. 56–57.

[121] NAI, HRR, vol. 825, 1782–1888, p. 204.

[122] *Legal and Expert Opinions on the Commercial Treaty of 1802, Hyderabad State*, Hyderabad: Government Central Press, 1932, p. 1.

[123] Regani gives a fascinating account of how residents had been pressuring the Nizam and prime ministers to carry out land revenue reforms and trade regulations, see her *Nizam–British Relations*.

all articles of merchandise of whatever kind imported into Hyderabad State and vice versa.[124] In short, the treaty abolished all other duties and fixed an import duty of 5 per cent. This treaty thus created a free flow of goods between the two territories.

Another important component of the treaty was the establishment of customs houses, and following it, the Nizam's government established *nakas* (customs houses) on all the main roads in 1808.[125] From the middle of the nineteenth century, market regulation took on a greater momentum. The British Indian government realised that unless customs houses were erected on the borders of the Nizam's territories it would be difficult to regulate the influx of goods into both the territories and to collect duties, and they put strong pressure on the Nizam to do this in the late 1840s.[126] After Salar Jung became the Prime Minister of Hyderabad State in 1853, the treaty was reviewed once again in 1855. Both the governments agreed to strict implementation of the norms of the treaty and to erect customs houses on the borders with the British territories, which the Nizam's government then did.[127] The British Indian government also set up number of its own customs check posts along the borders.[128]

Although the transit duty was abolished in Hyderabad State under the Treaty of 1802, the local customs officials continued to collect it. The situation was particularly bad in the jagirdar areas. In this regard, the British Indian government wrote a letter in 1873 to Salar Jung to stop the collection of transit duty in the state. In reply, Salar Jung asked the British Indian government to suggest a way of stopping this practice. They produced a scheme that Salar Jung then agreed to implement: according to this scheme, merchants carrying goods between British territories through Hyderabad State had to carry a receipt. The receipt would contain the details of the name of the trader, place of residence, description of goods, rate at which purchased, and place in the British

[124] *Legal and Expert Opinions on the Commercial Treaty*, p. 3.

[125] NAI, HRR, vol. 637, 1782–1888, p. 131.

[126] *Ibid.*, p. 227.

[127] NAI, HRR, Political, file no. 3/1860–65, p. 38. Customs posts were, for instance, set up at Nuldroog (Bombay road), Moogaputtum (near Aurangabad), Lingasugur (west of Raichur), Kodhad (Masulipatam road), Warapilly (Madras road) and Rajoorah (Nagpur and Berar road).

[128] NAI, HRR, Political, file no. 1/1862–65, p. 9.

territory to which they were being carried. This receipt was to be produced before the first *nakadar* of Hyderabad State at the border, and he would grant the merchant *dhakala* (proof receipt), retaining the old one with him. This dhakala was then produced again before the last nakadar in the Hyderabad territory, who would likewise examine the details of the receipt and permit the goods to pass, retaining the receipt. The two receipts would be then forwarded to the Central Customs House of the state for filing and future reference in case of disputes. If any goods were disposed of or purchased in the Hyderabad territories, they were taxed by the nakadars according to the existing rules, which was 5 per cent of the value of the commodity.[129]

Salar Jung was also instructed to change the internal market practices. A customs department similar to those in British India was established in 1860, and all non-official transit duties and other exactions were banned by law. Jagidars and maktadars were no longer allowed to collect duties from traders. They were compensated for this loss to their revenues with compensation in the form of *jagirs* (land grant), yielding an aggregate revenue of Rs 285,500 per annum, and also Rs 26,313 in cash.[130] Ironically, rather than being curtailed, malpractices and fraud, if anything, increased in the state thereafter.[131]

The new rules worked against the interests of the lambada caravanners, who were unable to handle the methodology and paperwork of the new regulations. Customs official would, for example, demand that they show dhakala for the goods they were carrying. Sometimes, even though they were carrying the dhakala, they were still stopped and harassed for mamul by the chowkidars. They were of course also made to stick to the route for which they had got passes, and had to pay fines if they deviated from it. Since they travelled with huge cattle herds, it was difficult for them to stick to a pre-determined route.[132]

[129] NAI, HRR, vol. 825, 1782–1888, pp. 204–07.

[130] *Report on the Administration of the Revenue and Allied Department of H. H. The Nizam's Government for 1324 F (1914–15)*, Hyderabad-Dn.: The Gladstone Press, 1916, pp. 46–49.

[131] NAI, HRR, Political, file no. 3/1860–65, p. 38.

[132] Andhra Pradesh State Archives, Nizam Custom Department (henceforth APSA, NCD), Inst. 13, List. 1, S. no. 1243, File no. R4/d226/1831, pp. 8–12; APSA, Nizam Revenue and Agricultural Department (henceforth NRAD), Inst 13, List. 1, S. no. 1172, File no. R4/d195/1927, pp. 1, 32.

If they breached the rules and changed their travelling routes, they were tracked down when they appeared at the village fairs or in the towns, and their property was confiscated.[133] This problem was more serious in the British territories as the market and customs regulations were strictly implemented there. In 1898, Lalchand and four other lambadas of Hyderabad State were travelling with 25 heads of cattle and six carts of jawar through Kandesh district (Berar) and were apprehended in Dholi *taluka* (sub-division of a district). They were brought before the local magistrate for not producing the customs duty receipts. The magistrate after consulting the resident at Hyderabad valued the cattle at Rs 445 and the grain at Rs 75. A fine was collected from them, which was double the actual customs duty.[134] Such experiences were common both in the British and the Nizam's territories. In fact, according to the agreement of 1865 between the two governments, salt was exempted from export and import duty, but a rate of Rs 2 per *pullah* or 120 seers was imposed as import duty, which was uniform throughout India, which was 25 per cent of the prevailing salt price. On top of this, the customs officials used to collect rahdari and transit duty from the lambadas. The lambadas were particularly harassed for dhakala at every customs chowki as they were continually on the move.[135] The criminal stigma attached to the lambadas during the nineteenth century in both British India and Hyderabad meant that their activities and movements were closely scrutinised, and they suffered correspondingly.

They suffered also as the trade at many local markets, fairs, jataras and uruses became unviable due to the new customs duties and trade regulations in the state.[136] The Administration Report of 1884–85 observed that 'the fairs suffered seriously from the vexatious transit dues extracted at almost every village on the road leading to Malligaon.'[137] As the local fairs wound up, trade activities shifted to small towns and

[133] APSA, NCD, Inst. 13, List. 1, S. No. 1243, File no. R4/d226/1831, p. 12.

[134] APSA, NRAD, Inst. 13, List. 1, S. no. 388, File no. R4/d46/1898, pp. 2–8.

[135] APSA, NRAD, Inst. 13, List. 1, S. no. 327, File no. R4/c3/1895, p. 9.

[136] *Ibid*., S. no. 167, File no. R4/d21/1886, pp. 2–3.

[137] *Report on the Administration of His Highness the Nizam's Dominions for the year 1294 F* (1884–95), Bombay: The Times of India Press, 1886, p. 246.

urban centres where conditions were less favourable for transporter-traders such as the lambadas.

The new market practices paved the way for the emergence of new trading communities that curbed the small-scale traders even further. As the Treaty opened free trade between British India and Hyderabad, many established traders from British India trained their sights on the stage.[138] Muslims, Parsis, Jains, Gujarati *baniya*s and *marwadi*s migrated to the state from western and northern India and gained control over its urban trade. The marwadis and Jain *vani*s even spread their market networks to the village level. By the end of the nineteenth century, the rural trade of each region of the state became gradually dominated by trading communities, notably the Jain vanis in the Marathwada region, the lingayat vanis in Karnataka, and komatis in Telangana.[139]

Some local peasant castes also started getting involved in trade activities; especially the richer strata that was already involved in money lending and other forms of exploitation of the subaltern peasantry. They gradually established themselves as merchants and traders. A recent study of the Warangal *suba* reveals how this phenomenon emerged with the new market practices. The *komati*s, who were in many cases rich landlords as well as village traders in Telangana, moved into wholesale trade, brokering and godown ownership. The list of prominent cotton dealers in Warangal included Siddamsetty Lingaya Besa, who was a komati, Pingali Venkat Janardhan Reddy and Pingali Pratap Reddy. All of them were from landowning communities, and the last-named was a powerful *deshmukh* (landlord) in Warangal taluka. The growth of new markets thus strengthened the hands of such landed communities.

[138] In the coastal districts of Madras Presidency, trade was largely organised and controlled by European and native businessmen. They maintained a number of agents and sub-agents in different towns to settle their transactions with the local merchants and to store their goods in warehouses till a purchaser was found. Trade was monopolised by a few merchants, especially in the ceded districts. These agents emerged eventually as independent traders and expanded their activities into Hyderabad. See Rao, *Economic History*, p. 40.

[139] *Imperial Gazetteer of India Provincial Series. Hyderabad State*, Calcutta: Superintendent of Government Printing, 1909, p. 43; also see Sheela Raj, *Mediaevalism to Modernism: Socio-Economic and Cultural History of Hyderabad, 1869–1911*, Bombay: Popular Prakashan, 1987, pp. 130–31.

Although still subordinated to the British cotton dealers, their power grew with the introduction of the railways.[140]

Hence, from the beginning of the nineteenth century, the new regulated market and colonial economy destroyed the long-established market relations in the state. It undermined the lambadas' trade network and their monopoly over the foodgrain trade. The new market forces made trading increasingly difficult and soon the lambadas were marginalised. To add to their woes, they were also badly affected by the severe trade depression of the period between about 1825 and 1845 that was caused to a large extent by the extortionate land taxes imposed by the British.[141] They were then marginalised even further by the introduction of the railways. The lambadas' oral literature suggests that this period was one of the most distressful in their history.

[140] V. Ramakrishna Reddy, *Economic History of Hyderabad State*, New Delhi: Gian Publishing House, 1987, p. 548.

[141] Bayly, *Rulers, Townsmen*, pp. 263–302.

Section III

WATER, IRRIGATION, RESOURCE STRATEGIES

Situating the Environment*

Settlement, Irrigation and Agriculture in Pre-colonial Rajasthan

Mayank Kumar**

Introduction

There can be multiple histories of a given time or place. These import to the given subject a certain richness. At times these histories converge, which, in turn, enable us to construct very different perspectives of the same temporal and spatial zones. Traditionally, studies of the medieval history of Rajasthan have focused largely on the glory of royalty on the one hand and the feudal culture on the other.[1] With the growing access of historians to administrative documentation maintained by various principalities, the little-explored aspects have come within the ambit of research. These documents have given deeper insights into the political, economic and social aspects

* Mayank Kumar, 'Situating the Environment: Settlement, Irrigation and Agriculture in Pre-colonial Rajasthan', in *Studies in History*, volume 24, issue number 2, July/December 2008, pp. 211–33.

** **Acknowledgements:** I would like to thank Dilbagh Singh, Ravindra Kumar, Yogesh Sharma, G. S. L. Devra and Shashank for insightful comments. An earlier draft of the article was read at 'Nature in the Raj: Colonial Environments and Indigenous Knowledge', a workshop organised by the Centre for Studies in Science Policy, School of Social Sciences, Jawaharlal Nehru University, New Delhi and the Centre for World Environmental History, University of Sussex, Sussex, 18–19 August 2006. I am grateful to the comments offered by participants, especially Mahesh Rangarajan, Meena Bhargava, Richard Grove, Rohan D'Souza and Biswamoy Pati.

[1] G. H. Ojha, *Rajasthan ka Ithihas*, Ajmer: Vaidika Yantralaya, 1938; Colonel James Tod, *Annals and Antiquities of Rajasthan or the Central and Western Rajput States of India*, 2 vols, repr., New Delhi: Motilal Banarsidass, 1971.

of the period. Literary evidence has further enriched these histories.[2] In these accounts, the dynamics of environment in the day-to-day functioning of the society has been at the margins. These works merely provide descriptions of the environmental features and support a tacit acceptance of the stereotype images of pre-colonial societies. Some of the writings have questioned the view wherein the state was attributed primarily the function of appropriator of agricultural surplus, being virtually unconcerned with the process of agricultural production.

An important concern for any agrarian society has been the availability of water in arid and semi-arid regions having unequal annual distribution of rainfall, and low water table that was often saline. Continuous availability of water was a primary concern, which requires a closer examination at the social and political levels. This will bring out the role and responses of social elites in the politics of water. It is all the more important as the requirements of subalterns translated into questions of survival even for the influential sections of society.[3] In the semi-arid and arid parts of Rajasthan, management of water

[2] Along with archival records, these accounts have relied on *khyal* literature as well. Satish Chandra (*Medieval India: Society, the Jagirdari Crisis and the Village*, New Delhi, Macmillan, 1982) began examination of the inner dynamics of village social structure with respect to revenue appropriation and distribution. S. P. Gupta (*The Agrarian System of Eastern Rajasthan c. 1650–1750*, New Delhi: Manohar, 1986) and Dilbagh Singh (*State, Landlords and Peasants*, New Delhi, Manohar, 1990) further developed it. They have examined the agrarian structure with reference to the state and social stratification, namely, the caste system. Scholars, such as G. D. Sharma (*Rajput Polity: A Study of Politics and Administration of the State of Marwar*, New Delhi: Manohar, 1977), G. S. L. Devra (*Rajasthan ki Prashashnik Vyavastha*, Bikaner: Dharti Prakashan, 1981), Dashratha Sharma (*Rajasthan through the Ages*, vol. 1, Bikaner: Rajasthan State Archives, 1966) and B. L. Bhadani (*Peasants, Artisans and Entrepreneurs: Economy of Marwar in the Seventeenth Century*, Jaipur: Rawat Publications, 1999) have examined the issue of agrarian relations to investigate the nature of medieval Rajput polity. Their central focus has been on revenue realisation and related issues. Nandita Prasad Sahai (*Politics of Patronage and Protest: The State, Society, and Artisans in Early Modern Rajasthan*, New Delhi: Oxford University Press, 2006) has examined another very important but often neglected component of social structure, that is, artisans.

[3] David Mosse (*Rule of Water: Statecraft, Ecology and Collective Action in South India*, New Delhi: Oxford University Press, 2003) has examined extensively

available during the three months of monsoon was, and still is, crucial for sustenance. Therefore, we witness several initiatives undertaken at the community level, alongside efforts made by the political functionaries in this regard.

This chapter examines the multiple dimensions associated with the management of the meagre water available in the region. The first section highlights various systems of water management developed and maintained at the local and individual levels alongside state initiatives on a larger scale for irrigation and potable purposes. Pre-colonial polities of Rajasthan were flexible in extracting revenue from a highly erratic, monsoon-dependent agricultural society. The next section details the administrative mechanisms evolved by states to address these concerns. It is argued here that the state had to ensure continuity of habitation by offering concessions and support to protect the revenue base. This was not an easy task for a state which had to rely on the support of the local ruling groups which in turn had their own interests to protect. Still, it can be argued that the state did have an active role in promoting agriculture and influencing agrarian relations. The nature of these interventions reveals that environmental factors, especially uncertainties, were important considerations.

Physical Settings and Vagaries of Monsoon

To examine the linkages between the environment, social structure, cultural values and the necessity for the state to monitor the various issues related to the appropriation and uses of water, it is necessary to delineate the physical features of the area of study. Geographically, Rajasthan is highly heterogeneous and, thus, difficult to classify as a single environmental entity. The Aravalli hills, running diagonally from the south-west to north-east, divides the region broadly into two natural divisions, namely, the north-western and the south-

---

the functioning of the tank system of irrigation as practised in the coastal regions of Tamil Nadu. It was an extensive network of tanks entertaining inter-village regulations and management. The management also required that due importance be accorded to caste considerations. However, we do not have such references for our region, perhaps due to the meagre rains.

eastern, the former being arid and the latter being semi-arid. About three-fifths of Rajputana lies to the north-west of this range of mountains and two-fifths to the south-east. Even these two regions are not homogeneous.[4]

As far as the general weather system is concerned, heat dominates[5] in the whole of Rajasthan except for some hilly parts. In the summer, the temperature rises above 40°C—the maximum occasionally reaching 50°C at several places in the desert areas.[6] As such, large parts of India may be termed as semi-arid, but the relative absence of perennial rivers further worsened the conditions in Rajasthan. Rivers like Looni, Chambal and Betwa are seasonal; no snow-fed rivers exist unlike in Punjab, north India or Assam. Rainfall also varies greatly depending upon the route of monsoons and other related factors. In the western section, that is, Jaisalmer, Bikaner and most of Marwar, the annual rainfall seldom averages more than 12 cm. Since the clouds have to pass extensive and heated tracts of sand before reaching these plains they lose much of their moisture upon the high ranges in Kathiawar and the nearer slopes of the Aravallis.[7] However, in south-west Rajasthan, which directly receives the monsoon with less intermediate evaporation, rainfall is much more abundant. In fact, at Mount Abu, it sometimes exceeds the level of 250 cm. Except in the south-west highlands of the Aravallis, rain is most abundant in south-east Rajasthan. Here, if the south-western rains fail, the south-eastern rains usually come to the rescue later in the season. Consequently, the region rarely

[4] *The Rajputana Gazetteer*, vol. 1, Calcutta, 1879, p. 20.

[5] On his way, Babur realised that north-east Rajasthan was extremely hot. See Zahir ud-din Muhammad Babur, *Babur-Nama*, trans. A. S. Beveridge, New Delhi: Oriental Books Reprint Corporation, 1970, p. 577; Abu'l Fazl and Jahangir placed it in the second of the seven categories in which the whole world was traditionally divided. The second category implied equitive heat and winter. See Abu'l Fazl, *Ain-i-Akbari*, vol. II, trans. Colonel H. S. Jarett, New Delhi: Oriental Books Reprint Corporation, 1978, p. 273; Jahangir, *Tuzuk-i-Jahangiri*, trans. H. Beveridge, New Delhi: Munshiram Manoharlal Publishers, 1968, pp. 340–41; Abdul Qadir Badauni, *Muntakhab-ut-Tawarikh*, vol. II, trans. W. H. Lowe, New Delhi: Renaissance Publishing House, 1986, p. 239.

[6] *The Rajputana Gazetteer*, 1879, p. 20.

[7] *Ibid*.

faces extreme droughts as in the case of the north-western tracts.[8] If the eastern winds are strong, they bring heavy rains from the Bay of Bengal, whereas if the south-west monsoon prevails, the rains are comparatively late and light. Sometimes a good bout of rainfall comes from both the seas, but this is rare. Rainfall varies substantially from year to year and even within a year.[9] Similar variations are visible in the natural vegetation as well. North-western Rajasthan is virtually devoid of any forest cover, with grass and bushes dominating the landscape;[10] the south-eastern part has good forest cover. In fact there are references of several *shikargahs* of the Mughal period.[11] Also, this semi-arid biogeographic zone was home to four large cats, namely, lions, tigers, cheetahs and leopards.[12]

Social Canvas

In medieval Rajasthan, 'the political and social orders were closely intertwined, integral parts of a continuum, webbed together through intricate ties'.[13] Society was characterised by the hierarchical order based on birth and hereditary duties and privileges.[14] This socio-

[8] *The Imperial Gazetteer of India*, vol. XXI, London, 1908, p. 92.

[9] *Arzdasht Asad Vadi* 1, 1695 vs./1638 CE; *Bhadva Sudi* 6, 1695 vs./1638 CE. Historical Section, Jaipur Records, Rajasthan State Archives, Bikaner. (henceforth, HS, JR, RSAB respectively).

[10] Irfan Habib, *An Atlas of the Mughal Empire*, New Delhi: Oxford University Press, 1992, Sheet no. 6, A & B; Irfan Habib, 'The Geographical Background', in Irfan Habib and Tapan Raychaudhuri, eds, *The Cambridge Economic History of India*, vol. 1, Cambridge: Cambridge University Press, 1982, pp. 1–13.

[11] *Arzdasht, Migsar Vadi* 13, 1697 vs./1640 CE; G. S. L. Devra, 'Desertification and Problem of Delimitation of Rajputana Desert during the Medieval Period', *Human Ecology*, Special Issue, no. 7, 1999, pp. 97–107.

[12] Some of the famous protected areas here are Ranthambore, Sariska, Mount Abu and Sitamata. See Divyabhanusinh, *The End of Trail: The Cheetah in India*, 2nd revised edition, New Delhi: Oxford University Press, 2001.

[13] Sahai, *Politics of Patronage and Protest*, p. 235.

[14] Dipankar Gupta, 'Continuous Hierarchies and Discrete Castes', in Dipankar Gupta, ed., *Social Stratification*, New Delhi: Oxford University Press, 1991, p. 141.

political stratification determined, among other things, access to natural resources of the different sections of society. It is important to note that pre-colonial polities had a strong element of kinship.[15] At the top in the hierarchy was the head of the clan, who had acquired territorial rights and established the principality; then came the *thikanadars* (nobles) who were his *saga* (blood relatives) brothers, sons, etc. The dominant sections of the village society comprised *jagirdars*, *bhomias* and *bohras* (assignment holders, right holders and merchant moneylenders respectively) who owed their status partly to hereditary superior proprietary rights in land and partly to their position in the revenue administration.[16] Bohras were the providers of borrowed capital for the ordinary peasantry. The rulers offered a certain amount of patronage and protection to bohras, as the political powers recognised their importance.[17]

The political apparatus ensured privileges not only for the wealthy and influential sections, but for the weaker sections of society as well. Ritual prescriptions were also at times bypassed because of the need for sufficient supply of labour to work the cultivable land available in abundance.[18] This was not a condition peculiar to Rajasthan as James C. Scott has argued. The key to successful statecraft was typically the ability to attract and hold a substantial productive population within a reasonable radius of the court. Given the relative sparseness of the population and the case of physical flight, the control of arable land was pointless unless there was a population to work it.[19]

Political power cannot be measured solely in terms of geographical expanse. The argument that 'a growing, productive population settled in the ambit of a monarch's capital was a more reliable indicator of a

[15] Sharma, *Rajput Polity*, pp. 1–58.

[16] Gupta, *The Agrarian System of Eastern Rajasthan*, pp. 134–43; Singh, *State, Landlords and Peasants*, pp. 42–50.

[17] Dilbagh Singh, 'The Role of the Mahajans in the Rural Economy in Eastern Rajasthan during the 18th Century', *Social Scientist* 2: 22, 1974, pp. 20–31.

[18] Sahai, *Politics of Patronage and Protest*, p. 127; also refer to footnote 12 on the same page.

[19] J. C. Scott, *Weapons of the Weak: Everyday Forms of Peasant Resistance*, New Haven: Yale University Press, 1985, p. 185. Sumit Guha has argued that Southeast Asia has always seen greater density since ancient times. See Sumit Guha, *Health and Population in South Asia from the Earliest Times to the Present*, New Delhi: Permanent Black, 2001.

kingdom's power than its physical extent'[20] seems to be closer to reality. There was a continuous attempt on the part of the ruling classes to increase the number of settlements in their territory.[21] For the same reason migrations during the years of environmental distress were actively contained by offering monetary assistance and fiscal concessions along with material support.[22] Even those who migrated were brought back with promises of concessions and remittances in revenue demand.[23]

Tracks of Evidence

Most of the evidence for this chapter is based upon the seventeenth- and eighteenth-centuries' archival and literary records of the Rajasthan states. The documents preserved at the state archives in Bikaner are significant as they provide an insight into the functioning of the state systems. At the same time, these documents also record the hardships faced by the peasants engaged in the process of agrarian production in pre-colonial times. Amongst the rich archival collection available for this region, we have used particularly the *arzdasht*s (correspondence between *pargana* officials and the central administration of Jaipur) and *sanad parwana bahi*[24] (royal directives issued to pargana officials in response to various complaints and representations received by the

[20] Locating the relative strength of a state between Occidental and Oriental regions, Scott has stressed that Southeast Asia has traditionally witnessed low population density and, therefore, the state actively guarded its human wealth. See J. C. Scott, *Seeing Like a State: How Certain Schemes to Improve the Human Condition Have Failed*, New Haven and London: Yale University Press, 1998, p. 185.

[21] *Arzdasht, Jeth Sudi* 4, 1742 vs./1685 CE; *Falgun Sudi* 2, 1743 vs./1686 CE.

[22] The pre-colonial kingdom thus rode a narrow path between a high level of taxes and coercive reactions that would precipitate wholesale flight. See Scott, *Seeing like a State*, p. 185.

[23] *Arzdasht, Magh Vadi* 9, 1765 vs./1708 CE; *Asad Sudi* 2, 1762 vs./1705 CE; *Jeth Vadi* 7, 1752 vs./1695 CE; *Chait Vadi* 3, 1752 vs./1695 CE.

[24] The subject matter of these documents (seventeenth–nineteenth centuries) ranges from routine complaints of undue collection of revenue by the state officials to the mutual disputes related even to the amount of water of a well to be shared among them.

court at Marwar, Jodhpur). Along with these, a seventeenth-century administrative treatise, *Marwar ra Pargana ri Vigat*,[25] written by Diwan Munhot Nainsi (1610–70) is used in this study. These documents provide details of administrative mechanisms from the pargana to the village level and throw light on the various environmental concerns of the medieval polities in the region. A careful reading helps us understand the relationship between the state, agricultural production and the peasantry.[26] This rich documentation enables us to reconstruct various indigenous methods of water management practised by cultivators to negotiate the vagaries of nature.[27]

Water Systems, Habitation and Irrigation

The general geophysical features of the region suggest that there were inadequate water resources, thus, leading to limited possibilities of human settlement and agriculture.[28] References in official documents of such settlements which perhaps did not contribute significantly to the state exchequer and their methods of water appropriation are, on the one hand, testimony to the concerns of the state and, on the other, point out how the land was put to constructive use. Beginning from village to the highest level, an intricate web of socio-political intermediaries characterised the nature of administrative apparatus. Therefore, it was possible for the political authorities to understand landscape at the very

[25] It is basically an administrative manual providing detailed description (seventeenth–nineteenth centuries) of the territory, soils, nature of habitation, agricultural types, nature and number of wells, other sources of water, caste and class composition of population, etc. Details of villages, which were deserted and later rehabilitated, also find mention here.

[26] However, these documents seem to be of limited use with respect to the examination of the role of lower classes/castes. It demands further research to examine such aspects.

[27] Mayank Kumar, 'Environment and Society in Medieval Rajasthan', Unpublished Ph.D. Thesis, Centre for Historical Studies, Jawaharlal Nehru University, New Delhi, 2001, pp. 241–306.

[28] 'The alternative route through Rajputana was in general sparsely inhabited …'; see W. H. Moreland, *India at the Death of Akbar: An Economic Study*, New Delhi: Reprints and Transpublications, 1962, p. 11.

local level. The decentralised nature of political authority facilitated prompt responses to ecological needs. The selection of sites for the construction of water systems and the methods of distribution and utilisation suggest that the administration had a fairly good idea about the landscape.

Various indigenous methods of water conservation were prevalent in the pre-colonial period, some of these are still in use today. Mostly practised at the individual level, some methods required larger community participation along with political support. Wherever feasible, the most important method to secure water was through wells. With the variations in the depth of the water table, the structures and functioning of the wells also changed. A lower water table necessitated deployment of water-lifting devices.[29]

Nainsi, the seventeenth-century court chronicler of Marwar, has described several types of wells, such as *chanch, kosita, dhimra* or *dhibra, kohar, kuwa* and *bera.*[30] These could be distinguished on the basis of their function and utility for irrigation and potable purposes. The character of the sub-soil, especially in the desert and mountainous parts of Rajasthan, renders the construction of wells or tanks labour intensive and, therefore, expensive. Limited prevalence of brick-lined wells suggests a high cost of construction.[31] Enumeration of earthen and brick-lined wells separately also testifies to the difference in their feasibility and, therefore, significance.[32] Detailed mapping of the wells in the Marwar region by the finance official (*diwan*) of the kingdom

[29] These devices differed according to the topography and the depth of the water table. The selection of a water-lifting device differed according to the nature of requirement, that is, potable or irrigation. The state was urged to provide loans to peasants in order to enable them to buy *lao-charas* (rope and leather) so that they could operate wells to irrigate crops. *Arzdasht, Asoj Vadi* 12, 1752 vs./1695 CE; *Asoj Vadi* 1, 1751 vs./1694 CE; *Kartik Vadi* 3, 1751 vs./1694 CE.

[30] Munhot Nainsi, *Marwar ra Pargana ri Vigat*, vol. III, Narain Singh Bhati, ed., Jodhpur: Rajasthan Oriental Research Institute, 1974, pp. 129–38.

[31] P. W. Powlett, *Gazetteer of Bikaner*, Calcutta, 1874, p. 90.

[32] Although brick-lined wells were comparatively costly, they had a longer life span and facilitated longer settlement. See Munhot Nainsi, *Marwar ra Pargana ri Vigat*, vol. II, Narain Singh Bhati, ed., Jodhpur: Rajasthan Oriental Research Institute, 1969 and Nainsi, *Marwar ra Pargana ri Vigat*, vol. III.

showed that the state was aware of the necessity of the sources of potable water for human settlements and for irrigation to ensure good agricultural production. There is evidence that political agencies also extended support for the maintenance of wells. In village Dhadahanriyo in pargana Jodhpur, a well which had became non-functional because it was covered with sand was restored with state support.[33]

Apart from digging wells, which was the most common method of rainwater harvesting, several other methods were used. One such method of water harvesting was variously known as *kuin* or *par* or *teen*. In appearance, it was similar to a well, but used a different mechanism to extract water which evolved to adapt to the particular environmental condition.[34] A kuin was not meant for round-the-clock availability of water[35] and was, therefore, relevant only for potable purposes. It was constructed in a few parts of Rajasthan, namely, Churu, Bikaner, Jaisalmer and Barmer. These regions had an underground layer of chalk or limestone which checked the percolating rainwater deep into the brackish water. The region being sandy, it was difficult to construct tanks because sand does not retain water. Geologically, the layer of limestone underneath ensured that rainwater remained potable and could be harnessed. It was more so because sand does not bind like soil and could retain water in it with restricted evaporation. Due to the low clay content and the dominance of sand content, it was difficult to construct conventional wells in the region as these could collapse. The kuin was a different kind of well in the sense that it did not depend upon underground water. In these kuin, water accumulated very slowly. The rainwater so retained by sand gradually percolated to the bed of the well and was usually not more than three earthen pots a day. The well took time to recharge and hence was used for drinking purposes only. The reference to these wells by Nainsi suggests the significance of these devices in continuance of human settlement.[36]

[33] Munhot Nainsi, *Marwar ra Pargana ri Vigat*, vol. I, Narain Singh Bhati, ed., Jodhpur: Rajasthan Oriental Research Institute, 1968, p. 296.

[34] Nainsi, *Marwar ra Pargana ri Vigat*, vol. II, p. 96.

[35] Nainsi, *Marwar ra Pargana ri Vigat*, vol. I, pp. 313, 327 and Nainsi, *Marwar ra Pargana ri Vigat*, vol. II, p. 96.

[36] Village Bhojavas of pargana Jodhpur had a *teen*, which could supply water only twice in a single day. See Nainsi, *Marwar ra Pargana ri Vigat*, vol. I, p. 313.

Similarly, the *bera* or *beri* is a variation of a well. Beras were constructed near a water body or on the dry bed of a water body.[37] It is extensively used till date in western Rajasthan as the rate of evaporation in the open ponds is high, but only nominal in the case of beras.[38] Similarly, another important manoeuvring of the environmental niche was practised to manufacture salt in the region. In most parts of this region of Rajasthan, groundwater was brackish, thereby providing the option to produce salt. In pargana Pokaran, there was a *rin*, a salt land. There is ample evidence that the water procured through beras was spread in the fields to dry up[39] and after several stages salt was obtained.

An important method of water conservation was the *kund* or *kundi*.[40] It was a mechanism based on the direct accumulation of rainwater in an underground brick-lined tank. It was primarily constructed for potable purposes in the north-western Rajasthan, where groundwater was either brackish or available at great depths or both.[41] Kundis were

[37] A. H. E. Boileau, *Personal Narrative Tour through the Western States of Rajwara in 1835 Comprising Beekaner, Jesulmer and Jodhpur*, Calcutta: Baptist Mission Press, 1837, pp. 33–34, 37, 49, 105.

[38] Village Godhali of pargana Phalodi had beri near seasonal streams. See Nainsi, *Marwar ra Pargana ri Vigat*, vol. II, p. 21. Village Naberi of pargana Phalodi sustained itself on the dry bed of a seasonal stream, which flows only during the rains. See Nainsi, *Marwar ra Pargana ri Vigat*, vol. II, p. 25. In village Bhabhelai of pargana Jodhpur, the groundwater was brackish. The village was without any source of potable water, but a beri constructed on a nearby tank sustained the habitation. See Nainsi, *Marwar ra Pargana ri Vigat*, vol. I, p. 263. Similarly, village Mahev in pargana Sojhat had a bera in the *talab* of the village and provided water when the talab dried up during the summers. See Nainsi, *Marwar ra Pargana ri Vigat*, vol. I, p. 457. In Jaisalmer the bed of a water tank, Amarsagar, sustained several beris. Amarsagar tends to retain water for only six to eight months and for the remaining part of the year these beris were the only source of water. See Anupam Misra, *Rajasthan ki Rajat ki Booden*, New Delhi: Gandhi Shanti Pratishthan, 1995, pp. 56–57.

[39] Nainsi, *Marwar ra Pargana ri Vigat*, vol. II, p. 36.

[40] The bigger form of the kundi is known as *tanka*. 'Occasionally where the wells were brackish, drinking water was supplied from *kundi*s or pits in which rainwater was collected and preserved, but in such small quantities as to be used only in an emergency'. See Boileau, *Personal Narrative Tour...*, pp. 13–14.

[41] *Ibid.*, p. 13.

constructed to cater both for domestic and public use.[42] It seems that the cost of construction of a kundi, even using the locally available material,[43] was beyond the means of the common man. Numerous epigraphs testify that usually these were constructed as a benevolent act by the wealthier section of society. The religious merit attached to the construction of water storage systems for the public seems to be one reason for the greater proliferation.[44] Followers of Islam too shared the notion of religious merit[45] associated with construction of water storage structures.

Tanks were built to harness rainwater, particularly in regions which had clay soils or hard ground surface that allowed limited percolation. We have a long list of tanks constructed by individuals, communities and rulers.[46] The 521 villages in the kingdom of Marwar had as many as 659

[42] The most important example of this mechanism is the Karor Pati Tanka of Jaigarh fort near Amber constructed in the fifteenth century. The parapet of the fort was used to collect rainwater and channelised to the tank located in the fort. See Rajiv Sharma and Syed Nadeem Ali Rezavi, 'Aspects of Hydraulic Engineering in Medieval Rajasthan: A Case Study of Water System in Jaigarh Fort', in Ahsan Jan Qaisar and Som Prakash Verma, eds, *Art and Culture: Felicitation Volume in Honour of Professor S. Nurul Hasan*, Jaipur: Publication Scheme, 1993, pp. 129–31.

[43] Details of materials used for construction of kundis are described in Misra, *Rajasthan ki Rajat ki Boondein*.

[44] Gopinath Sharma, *Rajasthan ke Itihas ke Srota* (in Hindi), 3rd edition, Jaipur: Rajasthan Hindi Granth Academy, 1995. For example, Pratapgarh stone inscription (946 CE), eulogy at the Lahar stepwell, Basangarh (Sirohi) (1042 CE), Indragarh Inscription 'A' (1683 CE), Bhuvana village stepwell inscription (1660 CE), eulogy at Trimukhi stepwell (1675 CE), eulogy at Gujar stepwell (1715 CE), Surtan stepwell inscription at Bedla village (1717 CE), Sagdogra stepwell inscription (1801 CE), Udaivav inscription at Dungarpur (1880 CE).

[45] Sharma, *Rajasthan ke Itihas ke Srota*. Sambhar stepwell inscription (1363 CE), Gaurishankar Tal, Narayana inscription (1437 CE), Makrana stepwell inscription (1651 CE), Bakalia inscription Dist. Nagore (1670 CE), etc.

[46] A few prominent tanks include the Udaisagar. See Munhot Nainsi, *Munhot Nainsi ri Khyat*, vol. I, Badri Prasad Sakariya, ed., Jodhpur: Rajasthan Oriental Research Institute, 1960, p. 30. Rajsummand constructed between 1605 and 1618 Vikram Samvat in Mewar. See Shyamal Das, *Vir Vinod*, vol. I, New Delhi: Motilal Banarsidass, 1886, pp. 401–49; Shyamal Das, *Vir Vinod*, vol. II, New Delhi: Motilal Banarsidass, 1886, p. 2204. Maharana

tanks. The highest number of tanks was in pargana Sojhat, followed by pargana Jaitaran and pargana Siwana.[47] Pokaran and Phalodi were also prominent for having a number of tanks. The environmental features of the region were suitable for tank irrigation. Lakes and dams were important sources of irrigation in the Mewat region.[48] However, of singular significance were the constructions of Garsisar in the fourteenth century and Jaitsar, a few generations later, in the vicinity of the city of Jaisalmer.[49] Selection of the site exemplifies a thorough knowledge of the local topography. The tank of Garsisar was constructed by erecting a *bund* (which means an embankment) to block the drainage of the catchment area created by the slopes of the hill on which the fort of Jaisalmer is located.[50]

Environment and Local Techniques of Water Conservation

In the absence of any snow-fed river and low water table on the one hand and the erratic nature of rainfall on the other, the only viable alternative to irrigation was to tap rainwater through artificial methods. People maximised the use of rainwater by evolving several indigenous methods of water harvesting for irrigation purposes like *khareen*, *rela*,

---

Raj Singh, the Sisodia Rajput ruler of the former princely state of Mewar (1662–80 CE), is credited with the construction of the Rajsummand, as engraved on the eulogy pasted on the embankment. See Neelima Vashishtha, 'A Note on the Sculptures of the Raj Samudra Lake in Rajasthan', in Ahsan Jan Qaisar and Som Prakash Verma, eds, *Art and Culture*, p. 111. Manasagar erected near Amber in the early eighteenth century. See I. A. Khan and Ravindra Kumar, 'The Mansagar Dam of Amber', in Aniruddha Roy and S. K. Bagchi, eds, *Ancient and Medieval Technologies in India*, New Delhi: Sundeep Prakashan, 1986, pp. 25–40.

[47] Bhadani, *Peasants, Artisans and Entrepreneurs*, p. 52.

[48] Babur extensively described the Kota lake and noticed its utility for irrigation. See Babur, *Babur-Nama*, p. 580. It lies partly in district Nuh, partly in Gurgaon, where the two tracts join at the foot of the Alwar hills.

[49] Lakhmi Chand, *Tawarikh Jaisalmer*, Jodhpur: Rajasthani Granthagar, repr. 1999, pp. 49–51.

[50] Lakhmi Chand, *Tawarikh Jaisalmer*, p. 41; *Jaisalmer ri Khyat*, in *Parampara*, parts 57–58, ed. Narain Singh Bhati, Jodhpur: Rajasthan Shodh Sansthan, n.d., pp. 46, 47, 58.

sewanj and *bahla*. These systems were essentially adapted to manage the run-off rainwater by directing it to the cultivation fields.

The practice of diverting rainwater to the fields for irrigational purposes (and diversion, in case of surplus, to neighbouring fields) is termed bahla. This was practised in regions with a clay soil and located usually in the vicinity of hills. The land thus inundated was also used for the cultivation of crops. In village Sanwarla, in pargana Seewano, the drainage caused by rains on the nearby hill was stored in the field and used for irrigation.[51] Village Motisaro in pargana Seewano could produce 200 *mun* of Sewanj wheat with the help of the bahla in the village.[52] In village Mahawan, in pargana Pokaran, the overflow of a *talab* was stored in a small tank nearby.[53] In Kharia Neeba Ro, in pargana Sojhat, the land was fertile due to the availability of run-off rainwater from the nearby hill.[54] Similarly, in the village of Peepaliyo Kapadiya Ro, in pargana Jaitaran, the overflow was used both for agricultural purposes and to irrigate the *jor*—the village pastureland.[55] On these lands crops could be sown, especially the ones which needed limited water.

The water conservation and canalising method, called rela, was a regional variation of the bahla and functioned on the same principle.[56] The bahla method essentially utilised rainwater, whereas rela used overflow from seasonal streams. This method, however, could not be used for irrigating fields other than those inundated by the flood water of the seasonal streams. Lands irrigated by the rela method were generally located along the banks of rivers.[57] These are generally natural depressions which were put into productive use by the village communities. The rela sustained *rabi/unali* crops. The village Peepaliyo Kapadiya Ro utilised the rela method of irrigation to generate revenue

[51] Nainsi, *Marwar ra Pargana ri Vigat*, vol. II, p. 238.

[52] Nainsi, *Marwar ra Pargana ri Vigat*, vol. II, p. 239. The *mun* was a unit for measuring weight, approximately equal to 51.63 lb. See Irfan Habib, *The Agrarian System of Mughal India 1556–1707*, Bombay: Asia Publishing House, 1963, pp. 367–79.

[53] Nainsi, *Marwar ra Pargana ri Vigat*, vol. II, p. 329.

[54] Nainsi, *Marwar ra Pargana ri Vigat*, vol. I, p. 427.

[55] Nainsi, *Marwar ra Pargana ri Vigat*, vol. II, p. 516.

[56] Nainsi, *Marwar ra Pargana ri Vigat*, vol. I, pp. 207, 454, 455, 515; Nainsi, *Marwar ra Pargana ri Vigat*, vol. II, pp. 145, 239, 329, 346.

[57] Nainsi, *Marwar ra Pargana ri Vigat*, vol. I, p. 353.

amounting to Rs 1,091 for the year 1658, Rs 2,250 for the year 1659, Rs 817 for the year 1662 and so on.[58] Villages Anandpur Khas and Lambia, in pargana Merta, produced a good quantity of wheat and gram due to the availability of rela-irrigated land.[59] It should be kept in mind that rela was the overflowed water spread over fields and the crop grown on this field was known as sewanj,[60] and these were rabi crops. Usually, wheat could be grown due to low requirement of moisture and necessity of limited watering. Generally, winter showers provided the requisite irrigation for the better growth of these crops.

Another method of rainwater harvesting for irrigational purposes practised in the Jaisalmer region was known as khareen.[61] It is generally believed that *paliwal* brahmins who settled in Jaisalmer during the early medieval times developed khareen. The land irrigated through khareens supported a rich harvest of wheat.[62] Khareens were basically bunds created in the path of *nullah* or seasonal streams. The height of the bund was kept low so as to retain only limited water which could dry up in two to three months. The bund also led to the deposition of silt, which ultimately formed the bed of the tank. This dried tank bed was used to produce rabi or unali crops. Another advantage of keeping the height of the bund low was that the overflowed water could be utilised to create another khareen downstream, which would enhance land available for cultivation.[63] It is vital to note that cultivation with the help of khareen could be achieved in the region adjoining the hills with a rocky surface, but it was irrelevant for the sandy soil.

A thorough and insightful understanding of the landscape on the part of the locals is visible in the conceptualisation of these water systems and their efforts to harness the given potential were amply supported

[58] *Ibid.*, p. 516.

[59] Nainsi, *Marwar ra Pargana ri Vigat*, vol. II, p. 120.

[60] See also Bhadani, *Peasants, Artisans and Entrepreneurs*, pp. 78–79 and Habib, *An Atlas of the Mughal Empire*, Sheet no. 6, A & B; Habib, 'The Geographical Background', pp. 1–13.

[61] Lakhmi Chand, *Tawarikh Jaisalmer*, p. 92.

[62] Misra, *Rajasthan ki Rajat ki Booden*, p. 62.

[63] Unlike the coastal regions of Tamil Nadu, except for continuous water tanks and their catchments in the Jaisalmer region and in the city of Jodhpur, the author is not aware of any such system in this region perhaps due to meagre rains. See Mosse, *Rule of Water*.

by the contemporary socio-political agencies. The ruling elites actively supported the development of such systems at local levels due to economic exigencies as it claimed a share in the enhanced agrarian production. The following section examines these interventions as manifestations of socio-political concerns pertaining to the environment.

Coping with Water Scarcity: State, Society and Agriculture

We have enough evidence to conclusively argue that the state extended financial and material support for the construction and maintenance of these water systems. Most of the investment was made to retain cultivators and thus to ensure the continuance of the settlements and its agricultural production. The ruling elite extended support for the construction of water bodies to encourage habitation in the region so as to enlarge the revenue for the principalities. The system of mapping the physical environment clearly manifested in the nature of interventions made by the political agencies to mitigate the vagaries of nature in general, and monsoons in particular, is very perceptive. The administrative apparatus kept a strict vigil over human settlements and continuously explored the favourable landscape to develop new possibilities for expanding habitation. The ruling elites continuously aspired to make landscape more congenial and cultivable, and with the emergence and consolidation of Mughal power in north India, we witness a definite step forward. The *Ain-i-Akbari*, the administrative manual compiled during the reign of Akbar by Abu'l Fazl, made a beginning in this direction. It was soon imitated by Munhot Nainsi in his *Marwar ra Pargana ri Vigat* for Marwar, which seems to have been replicated in the *arshatta* documents maintained by the Amber state.[64] These documents represent a kind of politico-administrative mechanism that laid ever-increasing stress on both physical and social mapping of the realm.[65] The administrative reorganisation of territory

[64] This trend received unprecedented impetus with the beginning of colonial rule, which started collection of census data, compilation of gazetteers and conducted anthropological studies from the nineteenth century.

[65] At another level, this kind of documentation was seen in European societies as part of their process of modernisation which was visible at the level

under Mughal rule, that is, the creation of *suba*, *sarkar* and pargana boundaries consolidated the administrative set-up of the principalities in this region.

Though apparently similar, administrative structures developed in Rajasthan differed broadly from their Mughal counterpart. In Rajasthan, the administrative apparatus was an extension of the larger social structure where, historically, power was shared among clan members. However, the efforts directed at strengthening the administrative

---

of abstract as well as realpolitik. It was the product of two processes working in synchrony. First, as Scott puts it:

> ... the process of documentation is best conceived as a strong, one might even say muscle bound, version of the self-confidence about scientific and technical progress, the expansion of production, the growing satisfaction of human needs, the mastery of nature (including the human nature), and above all, the rational design of social order commensurate with the scientific understanding of natural laws. It originated, of course, in the west, as a bye product of unprecedented progress in science and industry. (Scott, *Seeing like a State*, pp. 4–13)

second, in terms of:

> ... utilitarian discourse which replaces the term 'nature' with the term 'natural resources' focusing on those aspects of nature that can be appropriated for human use. A comparable logic extracts from a more generalised natural world those flora and fauna that are of utilitarian value (usually marketable commodities) and, in turn, reclassifies those species that compete with prey on or otherwise diminish the yields of the valued species. Thus, plants that are valued become 'crops', the species that compete with them are stigmatised as 'weeds' and the insects that ingest them are stigmatised as 'pests'. Thus, trees that are valued become 'timber', while species that compete with them become 'trash' trees or 'underbush'. The same logic applies to fauna. Highly valued animals become 'game' or 'livestock', while those animals that compete with or prey upon them become 'predators' or 'varmints'. (Scott, *Seeing like a State*, pp. 4–13)

However, we do not find any similarity of purpose between these two distinct experiences.

structure may be seen as a political strategy to negotiate the diverse social[66] and ecological factors.[67]

At the hub of the revenue administrative set-up was a diwan, who supervised the assessment and collection of land revenue from the parganas, each of which had its own officials. He had to keep himself well acquainted with the affairs of the principality, particularly with the activities of pargana officials such as the *amil, amin, chaudhari, qanungo* and *potadar*. They were his medium of contact with village officials such as the *patel, muqaddam* and *patwari*. The diwan also ensured that none of these officials exacted illegal or forbidden cesses or such as had been remitted by the king.[68] The records maintained at the office of the diwan provided the administrative map and social profile, along with a synopsis of the seasonal variations pertaining to the ecological conditions.

The appointments of pargana officials such as the amil, amin, potadar, chaudhari and qanungo were made by the diwan. The revenue collectors (amils) and assessors (amins) worked under the direct supervision and control of the diwan. He was required to see that all papers such as the rolls of *jama*,[69] *awarija*[70] and *yaddasht*[71] were sent to

[66] The nobles were being gradually but systematically converted into mere revenue assignees. See Dilbagh Singh, 'Regulating the Domestic: Notes on the Pre-colonial State and the Family', *Studies in History*, 19: 1, 2003, pp. 69–86.

[67] Water was the most determining factor for crop production. The supply of water by natural precipitation was meagre and sporadic, even while the climatically-induced requirement for water was very incessant. Irrigation thus, by necessity, becomes the primary concern for crop production in the whole region. Means of irrigation have always been very important yet very limited. The availability of irrigation facility made a difference in agricultural productivity. The long as well as short-term variations in the climate and rainfall ruined crops and dried up fodder, thus making living conditions difficult. In the absence of any snow-fed river on the one hand and the erratic nature of rainfall on the other, the only viable alternative for irrigation was to tap water through artificial methods of irrigation. See Singh, *State, Landlords and Peasants*, pp. 51–53 and Bhadani, *Peasants, Artisans and Entrepreneurs*, pp. 41–42.

[68] Gupta, *The Agrarian System of Eastern Rajasthan*, p. 164.

[69] The amount of revenue assessed on an individual or a village or larger area.

[70] Literally, *awarija* is a diary, a daybook, a ledger, a rough notebook, an abstract of receipt and disbursement.

[71] A kind of patta granted to the jagirdars at the time of grant of jagirs

him regularly. He issued instructions to the pargana officials for special measures regarding cultivation during the famine period.[72] Through such mechanisms, the polities of the region were able to read the variations in the ecological conditions and offered remedial measures.

As pargana was a fiscal as well as a territorial unit, amil and the amin, the land-revenue officers, were responsible for implementation of royal decrees and initiatives. The amil was entrusted with the task of ensuring cultivation on all arable land and the collection of the assessed revenue from the cultivated lands. He had to take the help of the *tappadar*, chaudhari, qanungo, *shahana* and other village officials. All letters of receipt and disbursement such as arshattas had to be duly signed by the amil and the amin. The amil was also enjoined to expand the areas under cultivation by getting more land, especially *ujar* land (waste land), brought under the plough. He was authorised to issue new *pattas* or *vat-vih* (share in the cultivation) at concessional rates, ranging from one-third to one-fourth, to those cultivators who brought *parat* (fallow) lands under the plough. He regularly communicated information regarding famines and other natural calamities that affected the cultivation, and implemented the measures prescribed by the diwan, such as making provisions for *hal* (plough), *bij* (seed) and *bail* (oxen), for the *palti* (cultivators).

Instructions were also issued by the amil to the chaudharis, qanungos, patels and patwaris to search for and rehabilitate the *raiyat* (peasant), who had abandoned cultivation due to famine, among other reasons, and to provide them with the facilities for cultivation. The amil and the amin were both responsible for the collection of *baqaya* (arrears) in their pargana; the diwan demanded *muchalka* (paper of agreement) from the amil and the amin of each pargana owning responsibility for realisation of the arrears. The amin's main duty was to verify the area of land under cultivation and to assess it according to the existing rules and regulations. He was required to send the assessment and other papers maintained in his office to the revenue department. He was required to inform higher authorities about the condition of rains in his pargana. The complex web of administrative functionaries ensured greater and deeper intervention by the state into the rural landscape.

---

by the rulers or at time of restoration of it. Some of these have also recorded the number of ploughs and bullocks per cultivator along with their castes.

[72] Gupta, *The Agrarian System of Eastern Rajasthan*, p. 165.

Two other important pargana officials concerned with the land revenue administration were the chaudhari and qanungo. The offices of the chaudhari and qanungo, though hereditary, were subject to renewal by the king. The hereditary nature of these offices further strengthens the contention that the political apparatus was an extension of kin relations. The chaudhari was supposed to render all help to the amin and the amil in the collection of revenue. He also served as a counter check against the qanungo. As regards the duties of the qanungo, he was required to collect detailed information about the land revenue and enter the same in the records under different heads. Both the chaudhari and the qanungo were required to prepare the *taqsim*[73] of different villages. These records were very significant for the political agencies as they contained all the relevant details with respect to the socio-ecological settings of villages. They provided an insight into the agrarian potential of a village, taking into consideration the role of environmental factors.

The shahana was perhaps the lowest official in the hierarchy of the revenue administration. The main duty of a shahana was to keep a watch on the crops when they were ready or when they were taken to the *kotha* (where the grain was stored for the purpose of sale) or to the *ganj* (grain market where it was brought for sale).[74]

Apart from the many officials appointed by the state, there were several others who were part of the political apparatus on account of their hereditary rights. Most of them were part of the local elite or the influential sections of society. Prominent among them was the patel who was normally a peasant and sometimes a petty *zamindar*. On the one hand, he was the natural leader of the villagers, and on the other he was the official headman (*mukhia*). He represented the raiyat and forwarded their grievances to the pargana officials. The patel was not paid directly from the treasury, but was entitled to some customary sources of income, such as the *inam muqaddami, bisondh* (one-twentieth part) and *dagli* (Rs 14 at the time of foundation of new village). He was given a *parwana* for *nankar*, a cash allowance or land in lieu of his

[73] *Taqsim* records carried the summary account of revenue in the pargana including its villages; the total area of a pargana, area of uncultivated land especially the area under *sir, kalar,* hill nullah, etc. It also included area under cultivation, the area under *punya-udik* (charitable land) along with harvest-wise revenue realisation. See Gupta, *The Agrarian System of Eastern Rajasthan*, p. 322.

[74] Gupta, *The Agrarian System of Eastern Rajasthan*, pp. 166–79.

services. Along with the patwari and the shahana, he would receive a coconut at the time of *batai* (sharing of the harvest).[75] The patwari was the village accountant who prepared the *khasra* papers as well as accounts of revenue collections. His papers were all in the local language. In the khasra papers, he was required to record the name of *asami* (cultivator), the area of his holding (*arz* and *tul*), the cultivable area, the name of the crops sown in *sayallu* (*kharif*) and *unhalu* (rabi) and the share of the raiyat and the diwan (state) in both the categories (*khalisa* and *jagiri*) of land.[76] Thus, it can be safely argued that the state concerned itself with various types of ecological facets and had penetrated deep into the intricacies of environmental landscape for its politico-administrative needs.

To maintain a regular supply of revenue, it was necessary to ensure continuous settlements and, therefore, availability of water. With the assistance of the administrative set-up, state not only appropriated greater revenue but it could also locate and promote water harvesting methods. In pargana Pahari, the failure of rabi crops was attributed to the lack of well irrigation in most of the villages in 1727.[77] Construction of wells was promoted in regions with a relatively high water table.[78] The state extended loans to the peasantry to dig new wells.[79] Regions with deep water tables opted for tanks,[80] which were also state financed.[81]

[75] A great deal of responsibility devolved upon the patel in times of famine. Together with the qanungo it was his duty to rehabilitate the village, to bring *paltis* (cultivators) back to their fields and to make provisions for seeds, bullocks and ploughs through loans either from the local bohra or from the *raj*. He was required to recommend grants of agricultural loans (*takai-taqavi*) by the state in times of distress. See Gupta, *The Agrarian System of Eastern Rajasthan*, pp. 183–84.

[76] *Ibid.*

[77] *Yaddashti, Pargana Pahari Mauza-Darobastki*, 1784 vs./1727 CE, HS, JR, RSAB.

[78] *Sanad Parwana Bahi*, No. 1, 1764 vs./1707 CE; Bhadani, *Peasants, Artisans and Entrepreneurs*, p. 49. Wherever wells occurred more frequently in c. 1660, the water table was nearer the surface.

[79] *Arzdasht, Kartik Sudi* 13, 1743 vs./1686 CE; *Bhadva Sudi* 11, 1743 vs./1686 CE; *Asad Sudi* 15, 1743 vs./1686 CE; *Sawan Sudi* 13, 1740 vs./1683 CE; *Asoj Vadi* 12, 1752 vs./1695 CE.

[80] Munhot Nainsi, *Mumhot Nainsi ri Khyat*, vol. II, Badri Prasad Sakariya, ed., Jodhpur: Rajasthan Oriental Research Institute, 1962, p. 25; Das, *Vir Vinod*, vol. I, pp. 117–18.

[81] *Sanad Parwana Bahi*, No. 2, 1765 vs./1708 CE.

Rulers also offered financial support for the repair of wells.[82] Those peasants who dug up new wells were also offered concessions in land revenue.[83] Similarly, the ruler was urged by the local revenue officials to provide loans to peasants in order to enable them to buy *lao-charas* (rope and leather) so that they could operate wells to irrigate crops.[84]

The governing class extended assistance by providing *taqavi* loans. Peasants could obtain a loan from a *mahajan* either against the mortgage of property or against *malzamini* (surety) furnished by the state representatives, that is, the zamindar, patel or jagirdar.[85] To protect the peasantry, the authorities imposed several restrictions on the functioning of the bohras. The general principle with regard to all loans—*tagai*,[86] as well as loans advanced by the bohras—was that debt incurred in the current harvest period should be recovered immediately after the harvest or within the fiscal year. Usually, arrears on old loans could only be demanded if the output of the current harvest was considered adequate. The ruling powers' attempt to discourage recovery of old loans by the bohras suggests that such loans were generally unrecoverable because peasants were left with very little surplus and consequently had to be written off.[87] In fact, the state actively discouraged the recovery of the older loans to curb emigration by the peasantry.

Moreover, the practice of the state standing surety for the loan sanctioned to the peasantry clearly indicates that rulers took an active part in the agricultural production. It is interesting to note that most of the evidence for such practices comes from locations that were agriculturally favourable and where water was the only limiting factor. It also strengthens Scott's argument that in Southeast Asia, retention of labour, in the form of human population, was the prime concern of the political powers.[88] A closer examination of the pre-colonial landscape amply substantiates our contention that large tracts of land were

[82] *Ibid.*, No. 8, 1768 vs./1711 CE.

[83] *Arzdasht*, *Kartik Vadi* 3, 1742 vs./1685 CE.

[84] *Arzdasht*, *Asoj Vadi* 12, 1752 vs./1695 CE; *Asoj Vadi* 1, 1751 vs./1684 CE; *Kartik Vadi* 3, 1751 vs./1684 CE.

[85] Singh, 'The Role of Mahajans...', p. 22.

[86] Irfan Habib, *The Agrarian System of Mughal India 1556–1707*, 2nd revised edition, New Delhi: Oxford University Press, 2002, p. 295.

[87] Singh, 'The Role of Mahajans...', p. 22.

[88] Scott, *Seeing like a State*, p. 185.

completely uncultivated and under forest cover. A cursory look at the contemporary evidence makes it clear that there was sparse habitation in the region. Similarly, extensive pervasiveness of diverse wildlife, as has been corroborated in numerous contemporary accounts, suggests a sparse agricultural landscape.[89]

Sparse agrarian landscape does not necessarily stem from a sparse population. Mere availability of land and peasantry was not sufficient to ensure growth of an agrarian economy and, for our region, possibilities of irrigation played a very significant role. A brief explanation of the agricultural pattern in Rajasthan will place us in a better position to appreciate the dependence on irrigation. In India, the production of two crops per annum has been a general norm, but in Rajasthan, due to the scarcity of water, at times it was difficult to raise even one crop in a year.[90] The long-term and short-term variations in the rainfall could ruin crops and dry up pasturage, thereby making sustenance difficult. For instance, in 1659 CE, it was reported to the authorities that in *qasba* Aaveri, rain measuring ten fingers fell on *Sawan Sudi* '11', so ploughing could be done only for four or five days.[91] Similarly, for pargana Amber in 1647 CE, it was reported by *purohit* Harsram that from *Sawan Vadi* '7' to Sawan Sudi '1', there was little rainfall in the pargana and because of strong winds, ploughing was not done and production fell sharply.[92] Such conditions often led to the migration of the peasantry to other favourable locations.[93]

The choice of the crops that were planted varied according to the changes in climatic patterns and flexibility in agricultural practices. If rains were delayed, peasants shifted to cultivation of *moth*, which had the shortest maturity period. In the year 1660 CE Amar Chand and Sahib

[89] Habib, *An Atlas of the Mughal Empire*, Sheet no. 6-B and 16–20.

[90] Nainsi, *Marwar ra Pargana ri Vigat*, vol. II, pp. 12, 36, 258–59.

[91] *Arzdasht, Sawan Sudi* 11, 1716 vs./1659 CE.

[92] *Arzdasht, Sawan Sudi* 1, 1704 vs./1647 CE.

[93] An arzdasht written by Mouji Ram (*Kati Sudi* 15, 1831 vs./1774 CE) informs the state about meagre rainfall leading to drought. This resulted in the migration of the peasantry. In his arzdasht (*Jeth Sudi* 1, 1819 vs./1762 CE), Lal Chand Dala Ram informs about migration of peasantry due to drought and the resultant decline in revenue collection. Similarly another arzdasht by Ajit Das Man Ram (*Chait Vadi* 3, 1809 vs./1752 CE), informs that villages were deserted.

Ram, *vakils* of the Amber court, noted that due to meagre and delayed rains till the month of *Bhadva*,[94] only moth could be sown.[95] Similarly, in the year 1638 CE, two other vakils, Ajitdas and Man Ram, noted that due to drought, not even moth could be sown.[96] The statement that 'not even *moth* could be grown' testifies to the fact that moth was the last crop that could be sown in case rains got delayed. In situations of delayed or insufficient rainfall, land would be ploughed extensively to make it soft and then seeds were sown. During the monsoon of 1641, Kanwar Pal and Bhopat Ram had informed the *Sarcar* that some parts of pargana Averi had received insufficient rain and that land was being worked upon to make that soft for sowing seeds.[97] In the next year (1642), Kanwar Pal and Hari Valabh Das supplied information that in pargana Averi, wherever the soil was hard, it needed at least one more shower, and wherever the rains had been insufficient, artificial irrigation by wells was arranged as well.[98]

A close look at the land revenue collection process and a reflection on agrarian production and output clearly establishes the correlation between investments made by the ruling elite in irrigation facilities to improve the agricultural potential.[99] The differences in land revenue rates also reflect the same phenomenon. Irrigated land was taxed at a higher rate compared to the non-irrigated land, which is a good indicator of states' concern to improve irrigation facilities. Local ecological as well as socio-political conditions provided ample opportunities to the political agencies to promote their agenda. While negotiating disputes over the use of water for irrigation purposes in the semi-arid climate, political powers promoted their own interests. Rulers not only arbitrated on the issue of ownership of means of

[94] Traditionally *Bhadva* is the third month of rainy season and usually the last one also.

[95] *Arzdasht, Bhadva Vadi* 7, 1717 vs./1660 CE); *Arzdasht, Sawan Sudi* 9, 1752 vs./1695 CE.

[96] *Arzdasht, Falgun Sudi* 11, 1695 vs./1638 CE.

[97] *Arzdasht, Kartik Sudi* 9, 1698 vs./1641 CE.

[98] *Arzdasht, Asoj Vadi* 7, 1699 vs./1642 CE.

[99] Bhadani, *Peasants, Artisans and Entrepreneurs*, pp. 34–80. Dominance of kharif crops also reflects greater dependence on source of irrigation for the crops grown during non-monsoon/rabi season.

irrigation, but also issued directives on the amount of water to be distributed.[100] At times, the state had to arbitrate between claimants over the use of irrigational devices.[101] It is interesting to note that the political authorities ensured an increase in the land to be cultivated by digging new wells in the villages.[102] The state's concern for optimal use of irrigation facilities is also evident from an important document pertaining to pargana Sojhat of Marwar state. Similar evidence is also available from eastern Rajasthan. An arzdasht written by Fateh Chand Ramji from pargana Malarna dated *Paus Vadi* 5, 1716 vs. (1659 CE) informs Raja Sawai Jaisingh that water from wells was being used for irrigation purposes as directed by the state.[103] Clearly, the state was aware of the ecological realities and, therefore, almost always attempted to maximise the utilisation of available water.

Concerns for potable water forced the state to punish any attempt at inappropriate or negative utilisation of scarce resources. In the year 1744, it was reported from village Dhamorki, pargana Khori, that the patel had diverted water from the river without permission. He was punished with a fine of Rs 20.[104] In 1768, another complaint recorded that Rupa Patel of village Raitoli breached the embankment of river to divert water for his use.[105] In another incident, the state punished a Gujar for diversion of water from other person's field in qasba Baswa in pargana Bahatri in 1717.[106] A similar dispute is reported in the same year from village Neembla in pargana Bahatri where Goverdhan Brahman fought with Tulani Brahman over the issue of diversion of water and hit him with an axe. Subsequently, he had to pay Rs 11 as fine.[107] The village pond was used not only for washing clothes and ancillary activities, but also by cattle for drinking purposes. Thus, any act of polluting it was punishable. We have evidence from qasba Malpura in pargana Malpura that in the year 1734, Lona, who was

[100] *Sanad Parwana Bahi*, No. 2, 1765 vs./1708 CE.
[101] *Sanad Parwana Bahi*, No. 8, 1768 vs./1711 CE.
[102] *Sanad Parwana Bahi*, No. 8, 1768 vs./1711 CE.
[103] *Arzdasht*, *Paus Vadi* 5, 1716 vs./1659 CE.
[104] *Arshatta*, *Pargana* Khori, 1801 vs./1744 CE.
[105] *Arshatta*, *Pargana* Dausa, 1825 vs./1768 CE.
[106] *Arshatta*, *Pargana* Bahatri, 1774 vs./1717 CE.
[107] *Arshatta*, *Pargana* Bahatri, 1774 vs./1717 CE.

a dyer, was fined Rs 5.50 for washing dyed cloth in the pond, and thereby polluting the water.[108]

In the agrarian economy of this semi-arid and arid region, failure of the monsoon very often led to drought. Excessive dependence on the rain and the relative absence of perennial rivers further compounded the problem in Rajasthan. Cultivation of a single crop per year in most of the arid parts of the region further reduced the possibilities of accumulation of grains and thereby squeezed the capacity of the peasantry to cope with the vagaries of the monsoon. These droughts could easily assume the form of a famine. Famines crucially hit the material base of the peasantry. The high cost of seed and manure and the drying up of wells restricted the ability to irrigate the land. In such conditions, ploughing could not be done,[109] thus affecting agricultural production in the region. Drought or infliction of crops, scarcity of fodder and drying of ponds also made the cattle vulnerable to adverse conditions[110] and affected the pastoral communities of the region. To mitigate these conditions of hardship, migration towards more favourable places such as Mathura, Gujarat and Malwa was extensively practised. However, the state made all possible efforts to stop the migration and often, attempted to persuade the peasantry to return by offering remissions in land revenue demand.

The vagaries of nature and related uncertainties about the quantum of agricultural production necessitated flexibility in the methods of governance. Under such conditions, a kind of flexibility was necessary whereby revenue collection was not calculated in rigid terms, rather, it was revised to correspond to the actual production generated. Possibility of cultivation of a particular crop in a season was determined by the timely onset of the rains; the choice of crop could vary according to the variations in the climatic patterns. Thus, flexibility had to be inbuilt in an agricultural economy where income or revenue differed from season to season due to cultivation of different crops with variance in yields.[111] Considerable leeway was granted by the state to the local elite in the assessment of revenue. Revenue was collected on the actual agricultural

[108] *Arshatta*, *Pargana* Malpura, 1791 vs./1734 CE.

[109] *Arzdasht*, *Paus Sudi* 4, 1704 vs./1647 CE.

[110] *Sanad Parwana Bahi*, *Falgun Vadi* 10, 1786 vs./1729 CE.

[111] Gupta, *The Agrarian System of Eastern Rajasthan*, pp. 38–73.

production on the verification issued by the important elements and by consensus in local society.[112] On the testimonies of the local elite about the actual environmental conditions in both the agricultural seasons, various remissions were granted. It also strengthened the superior status of intermediaries. Intermediaries were as such a part of the administrative apparatus and ensured two-way communication between the state and the peasantry.[113] These intermediaries safeguarded the interests of the state on the one hand and were the guardian of the peasantry on the other.[114] Intermediaries were dependent on the state for the continuance of *patta*s while they needed the peasantry to cultivate the land, thereby ascertaining the smooth process of agrarian and ancillary production. From the central authority to the cultivator, inherent flexibility can be observed with respect to agricultural production.[115]

Concluding Remarks

The list of issues associated with water management and its use is elaborate and can be traced at several levels. This chapter has examined the availability and utility of various indigenous water systems and their somewhat complex role in the societies' efforts at mitigating the vagaries of nature. None of these operated in isolation and it is incorrect to attribute apathy to the political formations in pre-colonial times. Certain level of flexibility is also evident in their approach to make

[112] *Ibid.*, p. 164.

[113] *Arzdasht, Magh Vadi* 9, 1765 vs./1708 CE; *Asad Sudi* 2, 1762 vs./1705 CE; *Jeth Vadi* 7, 1752 vs./1695 CE; *Chait Vadi* 3, 1752 vs./1695 CE. Habib, *The Agrarian System of Mughal India*, 2002, p. 295.

[114] Rohan D'Souza, 'Rigidity and the Affliction of Capitalist Property: Colonial Land Revenue and the Recasting of Nature', *Studies in History* 20: 2, 2004, p. 255. To demonstrate their superior social status and economic position, ruling elite practised the euphemisation of economic resources. '... what the wealthy did was to transmute a portion of their disproportionate economic means into forms of status, prestige and social control by means of acts they passed off as voluntary acts of generosity or charity. This social control was, of course, again convertible into labour services and hence again into material wealth'. See Scott, *Seeing like a State*, pp. 306–07.

[115] Singh, 'The Role of Mahajans...', p. 22.

optimum use of the erratic nature of the monsoon. The socio-political apparatus showed a continued capacity to accommodate the irregularity of rains and consequent fluctuations in the agricultural production. This adaptability was also seen within the structure of revenue administration which would allow space to accommodate the rights, privileges and demands of different sections of society in the wake of an almost regularly erratic monsoon and agriculture contingent on it.[116]

Unfortunately, we do not have sufficient evidence to comment on the use of water bodies by different castes given the nature of the prevalent caste system in pre-colonial Rajasthan.[117] An issue of significance is the role of Turkish rulers with respect to the caste system in general, and access to common water sources in particular. An inscription found at the stepwell at Makrana carries an official order by Mirza Ali Beg prohibiting the lower castes from collecting water from the well alongside the higher castes.[118] Did notions of purity and pollution attached with the sharing of the water bodies concern believers of other religions as well, especially Islam, is a question difficult to address at present. Moreover, water rights and related issues, such as who was the owner of the water, primarily groundwater, was it more important with respect to irrigation, did the state enjoy any kind of proprietary right over the water, are matters that have not been answered yet.

Equally important is the issue of the maintenance and upkeep of irrigation mechanisms. Without the active participation of various sections of society, it would have been a difficult task and more so in a 'caste-ridden' society. I would like to be slightly unconventional

[116] Both of them have argued that pre-colonial states functioned in the larger domain of traditions and in a social order which was defined exclusively by birth. See Sahai, *Politics of Patronage and Protest*, pp. 237–72. D'Souza, in his article, 'Rigidity and the Affliction of Capitalist Property', p. 255, has suggested that in pre-colonial socio-political formations euphemisation played an important role.

[117] Any description of Indian society is not without distinct references to the caste system. Fazl (*Ain-i-Akbari*), Nainsi (*Marwar ra Pargana ri Vigat*, vol. I, *Marwar ra Pargana ri Vigat*, vol. II, *Marwar ra Pargana ri Vigat*, vol. III) and Lakhmi Chand (*Tawarikh Jaisalmer*, both 1899 and 1999) while providing the administrative classification have based their descriptions on caste categories.

[118] Makrana stepwell inscription (1651 CE) cited in Sharma, *Rajasthan ke Itihas ke Srota*, p. 227.

and raise the question as to what extent can we term these systems as 'common' in the sense of 'common property resources'? It has already been pointed that the caste system dominated social realities, yet the premium on human labour may have necessitated deviations from the normative prescriptions of purity and pollution. Ecological settings of the region must have diluted strict adherence to caste segregation, and the notions of 'common property resources' therefore need reconsideration. Were they influenced by the peculiarity of the ecological niche? At another level, it also raises questions about the nature of regional polities of the period. Growing monetisation and commercialisation of economy, initially during the Mughal rule and later with the greater and deeper penetration of European commercial interests, may have forced modifications in the methods of governance. The beginning of this process can be traced in the nature and purpose of documentation of and negotiations between social and natural resources. Surely these are larger issues and only further exploration will provide insights into related socio-political realities.

Irrigation, State and Society in Pre-colonial India*

Tripta Wahi

I

The chapter proposes to look at water as a parallel component to land in agrarian production. It intends to explore the political economy of irrigation and water relations in agrarian production and their import both for the state and the society in pre-colonial India.

Irrigation is human intervention in water for purposes of producing bio-mass. At the outset it is important to state that irrigation is distinct from water reaching any land mass on its own leading to its tillage. This distinction is important as the two are invariably confused even in the best of scholarly works. For instance, there is a clear distinction between the *sailab* (inundation) reaching the riverine areas leading to crop production and channelisation of the sailab through human intervention to take that water to any specific area for agricultural production.

Irrigation has had vital implications both for the state and the society. We may begin by stating the obvious that for agricultural production there are two main natural resources required: water and land. Both are different in nature. While land is, by and large, a fixed entity,[1] water is in a constant flux. Further, while the land once cleared and developed can be regarded largely as a constant natural resource for agricultural production, water on the contrary is a constant variable and, thereby,

* Tripta Wahi, 'Irrigation, State and Society in Pre-Colonial India', Occasional Paper, in *History and Society*, New Series 36, New Delhi, Nehru Memorial Museum and Library, 2013, pp. 1–42.

[1] The rivers emanating from the Himalayas were prone to frequent changes in their courses and, therefore, the riverine belt of these rivers accordingly remained susceptible to fluctuations.

a factor of instability in agriculture. Variability of water produces a constantly fluctuating agricultural landscape;[2] stability in agriculture is in direct proportion to human intervention in water.[3] As the pre-modern states and societies within the Indian subcontinent depended primarily on the agrarian surplus, irrigation was at the core of their organisation and existence.

In this chapter, the theme 'Irrigation, State and Society in Pre-Colonial India' is being explored in relation to two modes of irrigation, namely, canals and wells. These two modes have been chosen as they represent almost two ends of the scale in irrigation. It may be borne in mind that water is available in different forms, namely, precipitation, surface water and underground water and the form of water availability defines the nature of irrigation. Different sources of water require different scales of organisation and resources, technologies, structures of organisation and forms of labour for creating irrigational mechanisms and also involve different systems of usage, distribution and control. These features stand out conspicuously when dealing with the two ends of the scale.

While canals deal with running surface water, usually huge volumes covering large areas, the wells are field-specific and their source is underground. The role of the state and the society individually and in interdependence and interaction is distinct in these two modes of irrigation. Yet there is commonality in terms of right to water, labour appropriation and socio-economic and political consequences both for the state and the society.

While the structure of organisation, form of labour, mechanism of usage, distribution and control emerge more clearly from the canal irrigation, rights to water stand out in sharp relief from the field-specific mode of irrigation, the wells, because contestation takes place at the ground reality level.

There is centrality of irrigation both for the agrarian production as also for the structures and mechanisms of appropriation of the surplus

[2] See, Tripta Wahi, 'Water Resources and the Agricultural Landscape: Pre-Colonial Punjab', in Indu Banga, ed, *Five Punjabi Centuries*, New Delhi: Manohar Publishers, 1997, pp. 280–82.

[3] *Ibid.*, p. 282.

produce. The state itself was deeply involved in creating and controlling irrigation facilities. Indeed the water rights defined the nature of the state.

II A

For primary data for the canals the chapter confines itself largely to the region covered by the British Punjab,[4] which encompassed more than two *suba*s of the Mughal Empire, namely, Lahore, Multan and parts of Delhi.

There was a great deal of irrigation from canals in the pre-colonial Punjab.[5] Indeed, there were canals all over the region, notwithstanding Babur's observation that there were no canals in Hindustan.[6] There is plenty of evidence to show that all rivers, streams and *nullah*s, perennial or seasonal, large or small, had canals. It can be safely asserted that topography and gradient permitting, canals were made on all running surface water. The British carried out an extensive survey of all the existing canals in the region with a view to working out their future course of action regarding canal irrigation. They even recorded remains of the abandoned canals. The exhaustive British survey forms the initial and main basis of evidence to work on the theme under discussion. However, invariably this evidence gets corroborated and supplemented by information from a variety of other sources from the earlier periods of the history of the region.

[4] It covered ten degrees of longitude and seven degrees of latitude, namely from 69.2°E to 79°E and from 27.4°N to 34.2°N.

[5] For a large-scale prevalence of canal irrigation in some other regions, see Irfan Habib, *The Agrarian System of Mughal India 1556–1707*, second revised edition, New Delhi: Oxford University Press, 1999, pp. 33, 38–39 and James Heitzman, *Gifts of Power: Lordship in an Early Indian State*, New Delhi: Oxford University Press, 1997, pp. 37–47. It is well known that there was extensive irrigation from canals in Kashmir. See Francois Bernier, *Travels in the Mogul Empire*, trans. Archibald Constable, a revised and improved edition based upon Irivine Brock's translation, New Delhi: Oxford University Press, 1972, pp. 396–97, 399; see also p. 454 for channels in Sindh.

[6] Babur, *Memoirs of Zehir-ed-Din Muhammed Babur*, Lucas King, ed., trans. John Leyden and William Erkine, London: Longman, 1921, II, pp. 205–06.

There were two types of canals, namely, perennial and seasonal. The perennial canals by definition had a constant supply of water whereas the seasonal canals worked for a few months in a year. The duration of the seasonal canals synchronised with the optimal flow of the source of running water.[7] The seasonal canals are described as inundation canals in the British sources and by modern historians, but to avoid confusion between the actual sailab and the channelisation during the period of optimal flow of rivers etc. of the water through human intervention,[8] it is preferable to call them seasonal canals. While the seasonal canals could be located anywhere on the river, they tended to be concentrated in the middling courses of the river. Their operations were directly linked to the fluctuating flows of the river. Thus, for instance, of the 15 main seasonal canals in the Multan district, the Diwanwah, Mahmudwah and the Bahawalwah on the Sutlej worked from April to October, while the Sardarwah and the Sultanwah were operational from April to November and the Kabulwah and the two Jumwah Doobabs worked from April to September. On the river Chenab, three canals were operational from April to October and four from April to September.[9] Operation timings of these and such other canals were linked to the volume of flow of the rivers/streams on which they were built. The volume of flow of the Indus river is at its lowest from October to March while in the pre-monsoon months their flow gains from the melting snow of the Himalayas[10] and the river water gets diverted by the *bunds* to the canals from where it is conducted inland. The seasonal canals existed equally on the non-

[7] For flows of the Indus rivers, see 'The Indus and its Tributaries', in *Mountains and Rivers (21st International Geographic Congress India, 1968 Inde)*, B. C. Law, ed., Calcutta: National Committee for Geography, 1968, pp. 351–52; see also Tripta Wahi, 'Water Resources and the Agricultural Landscape: Pre-Colonial Punjab', pp. 268–69.

[8] For such a confusion, see, for instance, H. C. Verma, *Harvesting Water and Rationalizaton of Agriculture in North Medieval India, 13–16 Centuries*, New Delhi: Anamika Publishers, 2001, p. 23.

[9] 'Canals of the Mooltan District', *Selections from the Public Correspondence of the Board of Administration for the Affairs of the Punjab*, Lahore, 1852, I, pp. 1–13.

[10] See, 'The Indus and its Tributaries', pp. 351–52 and Tripta Wahi, 'Water Resources and the Agricultural Landscape: Pre-Colonial Punjab', pp. 268–69.

perennial rivers, but they were more dependent on precipitation as these rivers did not originate in the snow-clad Himalayas. Numerous canals on the Ghaggar and the Saraswati are an instance in point.[11]

The perennial canals cover short or long distances. The constant supply of water was due to their head-works/embankments being usually located in places where rivers debouch from the hills. The submontane region with so many streams and nullahs, having a constant supply of water, was full of perennial canals; such canals covered short distances in between numerous streams. Thus, for instance, canals from the two perennial streams in the Dera Ghazi Khan region represent such a phenomenon.[12] Making canals covering longer distances required far greater technical expertise, management and resources. Many such canals are known to have been built in the Punjab several centuries prior to its annexation by the British. Two large canals were built by Firuz Shah Tughluq in the 1350s and at least one by Akbar and Shah Jahan each. All these canals subsequently fell into disuse and were reconstructed, extended and/or remodelled by the subsequent rulers at different points of time. Firuz Shah Tughluq built an extensive network of canals in the Sirhind and Hissar region, the dry belt in the Sutlej-Jamuna interfluve, and linked these to two perennial canals with one each from the rivers Sutlej and Jamuna.[13] As stated earlier, the Firuzi canals were rebuilt by Akbar; Shah Jahan too had them repaired and extended.[14] Shah Jahan also revived the Firuzi canal on the Jamuna and

[11] J. D. Cunningham, 'Report of the Irrigation of the Gugur and the Sursootee', *Selections from the Public Correspondence of the Administration for the Affairs of the Punjab*, 1854–55, II, No. XXIV, pp. 383–469.

[12] F. W. R. Fryer, *First Regular Settlement of The Dera Ghazi Khan District in the Derajat Division (1869–1874)*, Lahore, 1876, nos. 163–67, pp. 59–60. See also, B. R. Grover, 'The Extension of the Irrigation System and the Administration of the Canal Works in the Punjab during the Mughal Age, 1556–1707 A.D.', in *Land Rights, Landed Hierarchy and Village Community During the Mughal Age, Collected Works of Professor B.R. Grover*, ed. Amrita Grover, Anju Grover Chaudhary and J. C. Dua, New Delhi: Originals (an imprint of Low Publications), 2005, I, pp. 218–19, 225.

[13] Shams Siraj Afif, *Tarikh-i-Firuz Shahi*, the *Tughluq Kalin Bharat*, trans. S. A. A. Rizvi, New Delhi: Raj Kamal Prakashan, 2008, (reprint), II, pp. 74–75. Yahya ibn Sirhindi, *Tarikh-i-Mubarak Shahi*, *ibid.*, p. 199.

[14] Foreign-Political Department, 31st December 1847, Nos. 2351–52, p. 88, No. 5, National Archives of India (henceforth NAI), New Delhi.

brought it further to Delhi.[15] The Shah Nahr on the Ravi, with its head-works at Madhopur, was built during Shah Jahan's reign. Running through the upper Bari Doab, it brought water to Lahore.[16] Ali Mardan Khan also made a perennial canal on the river Tavi; it was brought from Sodhra to Ibrahimabad, a town that he had founded and named after his son.[17] There were four more perennial canals on the Ravi; they could well be branches of the same canal. They all started from Shahpur but had different destinations, namely, Lahore, Pathankot, Batala and Biar Pati Haibatpur.[18] There was also a perennial canal from a branch of the river Aik; this served the Sialkot region.[19] Seasonal or perennial, these canals often fell into disuse and were revived periodically. Shah Jahan's Shah Nahr was rebuilt by Ranjit Singh as the Hasli Canal,[20] while the Khanwah, attributed to Akbar's period, was revived by the British in the 1840s.[21]

The names of the canals invariably indicate their origin. To cite a few examples, the Bahawalwah canal goes back to the Daudpotras of Bahawalpur, Kabulwah presumably owed its origin to the Kabul rulers who held sway over Multan for some time, the Dewanwah was made by Dewan Sawun Mal and the Shah Nahr by the Mughal rulers.

The construction of canals required technical expertise, planning, knowledge of the river regimes, topography of the area and its gradient, besides financial outlay, organisation of labour and the actual construction. By any standards the making of a canal required well-established state structures to carry out the work. However, there were notable exceptions where canals were made by private individuals/groups of individuals or villages/communities.

[15] Sujan Rai Bhandari, *Khulasat-ut-Tawarikh*, Punjabi translation, Patiala: Punjabi University, 2000, p. 36; B. R. Grover, 'The Extension of the Irrigation System...', pp. 227–91; Abha Singh, 'Irrigating Haryana', in *Medieval India*, Irfan Habib, ed., New Delhi: Oxford University Press, 1992, pp. 57–58.

[16] *Ibid.*

[17] Sujan Rai Bhandari, *Khulasat-ut-Tawarikh*, p. 82.

[18] Irfan Habib, *The Agrarian System of Mughal India*, p. 36. Grover, 'The Extension of the Irrigation System...', p. 219.

[19] Foreign-Political Dept, 4–11 August 1849, Prog. no. 87, 'Canals in the Rachna Doab', no. 18., NAI, New Delhi; see also, Grover, *ibid.*, p. 220.

[20] R. Napier 's 'Report on the Shah Nahr or Hasli Canal', no. 18, NAI, New Delhi, pp. 39–48.

[21] Foreign-Political Dept, 21 February 1851, nos. 148–69, NAI, New Delhi.

In the British sources while there is a good deal of information on the maintenance of the canals, there is very little information on the processes of construction of the canals by the earlier rulers. This abundance of evidence on the maintenance of the canals is due to the fact that when the British assumed power in the region they also assumed the role of their predecessors in maintaining the canals. Till they established their own structures, they used the existing system, 'native' as they called it, with some improvisations. The near total absence of information on planning and construction of canals by their predecessors may be due to the fact that they constructed new canals entirely on their own without having to fall back on the past. Nevertheless their surveys of the existing and abandoned canals occasionally convey some good insights into the planning processes. Also, as scattered pieces of information spread over several centuries, from different sources and contexts, are pieced together and analysed, there emerges a fairly coherent picture of the structures and processes of canal-making in the pre-colonial times.

In the Mughal times, improvement and expansion of agriculture was the responsibility of the *Sipah Salar.*[22] This presumed expansion of irrigational facilities. The *Subahdar* was, in any event, the custodian of water-related structures.[23] Decision to make canals presumably could emanate from him. However, it was Akbar who took the decision to rebuild Firuz Shah's canal.[24] This may be due to the fact that the canal fell within the *suba* of Delhi and that the *sarkar* of Hissar had been conferred on his very young son Salim. In any event, the long perennial canals in view of the hugeness of the project, must assume decision-making by the rulers themselves. However, relatively smaller canals are known to have been built by powerful individuals/groups/communities/groups of villages. Numerous examples of such canals are available from the districts of Dera Ghazi Khan, Bannu and Muzaffargarh.[25] In fact, the region had water lords.

[22] Abu'l Fazl, *Ain-i-Akbari*, trans. H. Blochmann, Calcutta: The Asiatic Society of Bengal, 2010 (reprint), II, *Ain I*, p. 39.

[23] *Ibid.*

[24] Lieut. Yule, 'A Canal Act of Emperor Akbar with Some Notes and Remarks on the History of the Western Jumna Canal', *Journal of the Asiatic Society of Bengal* XV: 171, Calcutta, 1846, pp. 214–25.

[25] See, for instance, Fryer, *First Regular Settlement of The Dera Ghazi Khan District...*, p. 59, no. 165; Edward O'Brien, *Land Revenue Settlement of the*

Rebuilding a canal may or may not involve any survey, but making a new canal would necessarily need reviewing topographical and gradient surveys. Such an exercise would involve experts to work out where to situate the head-works, particularly of the perennial canals and also the route to be followed in terms of feasibility and desirability; local knowledge would be an integral part of the entire exercise. In the 1840s when the British were carrying out a survey with a view to supplying water to the eastern Majha region, they discovered that they had been anticipated on the 'very best line of country' and 'in the most favourable direction, by the traces of an old canal'.[26] The opinion that the old canal followed 'the very best line of country' indicates that the survey for the decayed canal had been carefully carried out. Any lacunae in the survey could lead to catastrophe as it happened in the case of the *Nahr-i-Bihisht* during Shah Jahan's reign.[27]

The work of making a canal involved several agencies and structures at multiple levels. It is evident from Akbar's *sanad* concerning the reconstruction of Firuz Shah's canal on the Jamuna that a skilled *maimar* (architect/mason), the superintendent of construction and the *faujdars*, all worked in unison to make the canal.[28] Such coordinated activity of various agencies and structures of the state at multiple levels would be essential for any canal-making exercise, even seasonal canals. For seasonal canals small political units/chiefdoms could undertake this activity, but much bigger states and extremely well-coordinated and cohesive activity in different places at different levels was a given prerequisite for the canal system that was of the magnitude of the canals of the fourteenth century or of the sixteenth–seventeenth centuries.

In a huge network of canals drawing water both from the Sutlej and the Jamuna and linking it up with the entire Ghaggar system of rivers, Firuz Shah Tughluq himself seems to have organised this huge undertaking. He was keen to get a constant supply of water in the region of deficient precipitation without a perennial river. It is quite remarkable that prior to his decision to draw water from the rivers

---

Muzaffargarh District, Lahore 1882, pp. 13–17, S. S. Thorburn, *First Regular Settlement of the Bannu District*, Lahore, 1879, pp. 94–98.

[26] Foreign-Political Dept., 4–11 August 1849, R. P. Napier's, 'Report on the Shah Nahr or Husli Canal', no. 90, NAI, New Delhi.

[27] Abha Singh, 'Irrigating Haryana', p. 59.

[28] Lieut. Yule, 'A Canal Act of Emperor Akbar...', pp. 214–17.

Sutlej and Jamuna, Firuz Shah Tughluq had attempted to cut through a mountain to ensure a constantly-flowing stream in the region.[29] He did not succeed in his enterprise. This was followed by drawing water from perennial rivers on two opposite sides of the region. This huge exercise involved big labour. While Akbar, on revisiting the canal, could mobilise local labour everywhere,[30] this could not have been possible during Firuz Shah's period as much of the region was not under cultivation during this period and, therefore, sparsely populated.[31] Firuz Shah, it may be safely conjectured, used his own slave labour of which he was the master. As a matter of policy he was collecting slaves and training them in different spheres.[32]

Perennial canals covering long distances were far more difficult and complex to make than the seasonal or short distance perennial canals. Perennial canals covering long distances had invariably to negotiate difficult terrain and opposing streams and torrents on the way. Thus, for instance, the Shah Nahr had to cross the torrents of Chakki and Jena on the way.[33] Technical aspects and the execution of the project could only have been handled by the state.

Constant maintenance and vigil of the canals, including regulation of water supply, were essential components of the canal system. For both types of canals annual desilting was essential; the Himalayan rivers, because of the poor rock formation of the mountains, always carried heavy sediment with them. This affected both the head-works[34] and the canals as well as the sub-channels. So a systematic annual de-silting

[29] Sirhindi, *Tarikh-i-Mubarakshahi*, pp. 201–2.

[30] Yule, 'A Canal Act of Emperor Akbar...', pp. 216–17.

[31] From all sources cited regarding Firuz Shah's canals, this is the picture that emerges on the point under discussion.

[32] Afif, *Tarikh-i-Firuz Shahi*, pp. 112–14. Sirhindi, *Tarikh-i-Mubarak Shahi*, p. 206. Due to a huge increase in the number of slaves, the department of slaves (*Diwan-i- Bandagan*) was separated from the *Diwan-i-Vizarat*, of which it was originally a part. See R. C. Jauhri, *Firoz Tughluq* (1351–1388), Jalandhar: ABS Publications, 1990, pp.126–27.

[33] Foreign-Political Dept, 4–17 August 1849, no. 88, 'Canals in the Baree Doab', para 6, NAI, New Delhi.

[34] Foreign-Political Dept, 28 February 1851, pp. 1–3, Annexure: 'Detailed Statement of the Canals'. In 1850, four out of 10 canals of the Dera Ghazi Khan district required new 'mouths'.

was the norm. This was undertaken during the winter months when the flow of water in the rivers was low. Even during the operational season, a constant vigil was maintained on the canals for any repair that might be needed in case any *bund* got damaged. We may keep it in mind that the bunds were semi-permanent structures and, therefore, vulnerable to damage with a strong and heavy flow. The stoppage of breaches required constant watch.

On the Multan canals the annual repair, clearance, stoppage of breaches and all other expenses were borne by the 'public' and this was so on almost all canals when the British assumed power.[35] While the labour was provided by the villages using the canal water, the management was done by the state agencies. Within the villages using the canal water, work of desilting was carried out locally, but for the mouth the labour was requisitioned from villages.[36] The same norm was applicable to the main channel. However, where the canals were the property of private individuals/groups etc., their maintenance was also undertaken by the self-same agencies.

For an overview of the kind of establishment that was needed for the maintenance of the canals and the regulation of water supply, it may be noted that the British, after considerably reducing the establishment, still had 197 persons for the Hasli canal, among whom 130 persons were *beldars*.[37] In the mid-eighteenth century the British had learnt that on the extended Jamuna canal reaching Delhi, a *daroga* was stationed at every 3–4 *kos* for purposes of police and the ready repair of accidents and he had peons and beldars under him.[38] One thousand armed peons and 500 horses were supposedly maintained on the establishment.[39] Under Ranjit Singh even army contingents were assigned to keep vigil on the Hasli.[40] That a very close watch was maintained on the use of canal water is apparent from the fact that

[35] 'Mooltan Canals', p. 3

[36] *Ibid.*, pp. 3–4.

[37] Foreign-Political Dept, 4–11 August 1849, 'Canals of the Baree Doab'. Letter from Secretary to the Board of Administration to H. M. Elliot, Secretary to the Govt. of India, 'Note', NAI, New Delhi.

[38] Yule, 'A Canal Act of Emperor Akbar...', p. 222.

[39] *Ibid.*

[40] J. S. Grewal and Indu Banga, *Civil and Military Affairs of Maharaja Ranjit Singh*, Amritsar: Guru Nanak Dev University, 1987, Document no. 109.

in 1732 an order had to be issued to the *darogas* of the *Shah Nahr*[41] to restrain their *gumashtas* from charging *naharana* from the village Talibabad in the Batala *pargana*. Obviously, the gumashtas were not aware of the nature of the grant and they had started demanding the naharana when the new *muzarain* were settled in the village and the canal water had begun to be used. Hence, the canal establishment was truly functional at the ground level.

There was a state establishment for the allocation of water and the *mir-i-ab* (chief of water) was in charge of it. The *Ain-i-Akbari* tells us that the *kotwal* in the Mughal Empire was the appointing authority for the mir-i-ab.[42] However, Akbar, after virtually reconstructing Firuz Shah Tughluq's canal, conferred the title of mir-i-abs on Muhammad Khan Tarkhan, the superintendent of the work 'from first to last'.[43] He was presumably made the mir-i-ab for the entire canal while the appointments made by the kotwals would be at a lower level. The British records show several mir- i-abs on a single canal.[44] The office holder was supposed to be a person of integrity who would allocate water fairly to all.[45]

Firuz Shah Tugluq also had an establishment to oversee the reach of the water of the canals that were made by him. Some *maliks* were especially designated for that task.[46] Afif's own father and uncle had held this duty.[47] It may be safely assumed that the purpose of such an establishment was to regulate the supply of water with the intent of raising accurate naharana. Indeed, such establishments continued through the entire pre-British period in the Punjab. Ranjit Singh, as noted above, even sent army contingents to police the Hasli canal. Where the canals were built by the non-state agencies, the distribution and supply of water was regulated by the owners/ builders of the canals; the entire system of distribution was worked

[41] B. N. Goswamy and J. S. Grewal, *The Mughal and Sikh Rulers and the Vaishnavas of Pindori: A Historical Interpretation of 52 Persian Documents*, Simla: Indian Institute of Advanced Study, 1969, Document no. III, pp. 94–95.

[42] *Ain-i-Akbari,* II, *Ain* IV, p. 45.

[43] Yule, 'A Canal Act of Emperor Akbar...', p. 25.

[44] See, for instance, 'Mooltan Canals', p. 4.

[45] Yule, 'A Canal Act of Emperor Akbar...', p. 217; *Ain-i-Akbari*, II, p. 45.

[46] Afif, *Tarikh-i-Firuz Shahi*, pp. 75–76.

[47] *Ibid.*, p. 76.

out to its most minute detail and the British were struck by its accuracy and equity.[48]

II B

The making of canals found reverberations in the changing social impulse of the region within the reach of the canals. They changed the social dynamics of the region that they covered. The impact of the canals both on the state and society was immediate and multi-dimensional. It was reflected in the pattern of economic activity encompassing agriculture, horticulture and trade. It got physically reflected in the construction of buildings and beautification of the existing towns as also in the establishment of new towns. Indeed, the entire canal-making exercise was symbiotically linked to the establishment of new towns, centres of administrative power and control.

The canals stabilised,[49] expanded and diversified agricultural production. They gave impetus to the production of cash crops and orchards flourished everywhere the canals reached. In and around Hissar-Firuza where only one crop used to be grown, the network of canals led to the growing of two crops.[50] Both crops began to yield abundant produce and a variety of fruits and flowers began to be produced. Orchards and gardens were made on a large scale and they were made both by the Sultan and the nobles. Similarly Fathbad was established with a network of canals. The network of canals along with the newly created centres of power transformed the entire region. Agriculture improved dramatically both by way of variety and also production of high quality cash crops, besides expansion of the area under cultivation. The region began to yield such huge revenues that Firuz Shah Tughluq deemed it appropriate to convene an assembly of the Muslim learned and the devout on the issue of the revenues

[48] See, for instance, Thorburn, *First Regular Settlement of the Bannu District*, pp. 98–101.

[49] Stability in agriculture was in direct proportion to human intervention in water. See Tripta Wahi, 'Water Resources and the Agricultural Landscape: Pre-Colonial Punjab', p. 282.

[50] Afif, *Tarikh-i-Firuz Shahi*, pp. 74–75.

being generated by the canals.[51] The assembled personages gave the opinion that since the canals had been built entirely with the labour and expenditure of the Sultan, he was entitled to *haqq-i-shurb*, that is, one-tenth of the revenue so generated. Similarly, he founded and brought numerous villages in his *imlaq* (personal property) by bringing them under cultivation.[52] The revenues collected from the naharana and the imlaq went to his personal treasury as distinct from the *baitulmal* (public treasury).[53] Similarly, the founding of Ibrahimabad[54] with a perennial canal from the river Tavi by Ali Mardan Khan during Shah Jahan's reign translated itself into an analogous transformation of the area. He made in it a garden rivalling the Shalimar garden of Lahore, constructed numerous buildings and patronised men of learning. The place became famous as a centre of learning and good handwriting.[55] There was a substantial increase in the population of the area and Shah Jahan gave in *inam* a village to Ali Mardan Khan and his family for the maintenance of the canal.[56]

When the British took over the Sikh possessions, they found that they could not substantially repair or enlarge the Hasli canal as cultivation extended right up to the brink of the canal.[57] With the extension to Delhi of the Firuzi canal on the river Jamuna under Shah Jahan, channels reached innumerable villages around the capital city of Delhi and with these channels orchards came up all over the region. We may also recall that Akbar had fruit trees planted all along the banks of the reconstructed Firuzi canal.[58] Canals defined the landscape, the social habits and the milieu through prosperity and patronage. The canals, and for that matter all irrigational facilities, also encompassed within

[51] *Ibid.*, p. 75.

[52] *Ibid.*

[53] *Ibid.*

[54] Sujan Rai Bhandari, *Khulasat-ut-Tawarikh*, p. 82.

[55] *Ibid.* See also Bute Shah, *Punjab di Geographical Tawarikh (Geographical Description of The Punjab,* in Panjabi), translated from the Persian of Bute Shah by Munshi Bahlol in 1850, reprint, Chandigarh: Lok Geet Prakashan, 2007, p. 113.

[56] *Ibid.*

[57] Foreign-Political Dept., 31 December 1847, nos. 2351–52, 'Regarding the Revenues Obtained at Present by Irrigation', pp. 42–43.

[58] Yule, 'A Canal Act of Emperor Akbar...', p. 217.

their ambit of impact religious institutions and religious ambience of the region. With the revenues generated by the canal system, Firuz Shah Tughluq bestowed extensive patronage on the Muslim divines and learned men.[59] It is a distinct possibility that the family of the *pirzadahas* of Dhatrat, from whom the document 'A Canal Act of Akbar' was procured in the nineteenth century, goes back to Firuz Shah's patronage.[60]

The seasonal canals during their dry period would become easy footpaths for the people to travel by instead of having to traverse thorny and bushy wayfares.[61] Further, most canals were navigable for shorter or longer distances. It is remarkable that in the Multan region 13 out of the 15 main seasonal canals were navigable.[62] Akbar's clear instructions for the restructuring of the Firuzi canal to Hissar Firuza mention that the canal must be navigable; small boats could ply even on very small canals. These canals thus provided a network of communications to various regions and sub-regions and connected them with larger networks of communications. With the expansion of agrarian production and especially of cash crops, increase in trade would be a natural corollary. The network of canals by being navigable would facilitate trading activity. Trade too would benefit both the state and certain sections of society. Most of all, canals provided some safeguard against famines.

II C

The canals generated huge revenues for the state. Firuz Shah's personal treasury getting filled with immense wealth accruing from the canals directly or indirectly has been noted above. Increase in agrarian production and greater emphasis on the production of cash crops inevitably enhanced the state revenues and gave impetus to trade. Also there was naharana on the canal water. Increase in revenues came

[59] Afif, *Tarikh-i-Firuz Shahi*, p. 75. See also *Futuhat-i-Firuz Shahi*, p. 334.

[60] It may be noted that Dhatrat finds a mention in the context of the increased revenues from the region. See Afif, *ibid.*, p. 75.

[61] 'Mooltan Canals'. See also 'Dera Ghazi Khan Canals'.

[62] *Ibid.*, Annexure.

not merely from substantially enhanced agrarian production and the naharana, but also through the ferry tax. With canals being navigable, boats would be a constant source of income to the state. Increase in trade too would generate additional revenues through taxation. It is amply evident that there was a close and multi-dimensional relationship between the state, society and the canals.

The relationship between the state, society and the canals is best summed up in the preamble to Akbar's *sanad* regarding rebuilding of the Firuzi canal. We may let it speak for itself:

> My government is a tree, the roots of which are firm in the earth. In acknowledgement of God's mercy in establishing this great empire, my desire, purer than water, is to supply the wants of the poor; and the water of life in my heart is larger than the sea, with the wish to dispense benefits and to leave permanent marks of the greatness of my Empire, by digging canals, and founding cities, by which too the revenues of the Empire will be increased.
>
> *God says, sow a grain, and reap sevenfold. My desire is to reap one-hundred fold that my crown may become wealthy, and the zamindars may obtain double returns.*[63]

Remaking the canal and making water available is the 'best purpose to which my wealth can be applied'. Akbar goes on to add:

> For God has said, from water all things were made. I consequently ordain, that this jungle, in which subsistence is obtained with thirst, be converted into a place of comport, free from the evil...
>
> *'Behold the power of God, how he brings to life land that was dead'.*
>
> Truly a canal is opened...
>
> He (Akbar) is such a king, that from the canal of his liberality, the garden of the world is green all the year round...[64]

Akbar acquired name, fame, resources and a stable social base in the region. He was also concurrently hoping for a reward after death as

[63] Yule, 'A Canal Act of Akbar...', pp. 213–14. Italics mine.

[64] *Ibid.*, pp. 214–15.

is stated in the preamble, 'The seeds sown in this world are reaped in the next'.[65] This perennial canal was extended further to Delhi during Shah Jahan's reign. It reached thousands of villages around the capital. Except for the 'reward in the next world', history is testimony to the fact that the canal yielded all the multi-dimensional results anticipated by Akbar. According to a proverbial expression current in Delhi in the mid-eighteenth century, the net revenue from these canals was reckoned equal to the maintenance of 12,000 horses.[66]

II D

The material benefits accruing to the state and different sections of society from the making, remaking and maintenance of canals came from the labour provided mostly by the villages on the canal routes. With the exception of Firuz Shah Tughluq, who seems to have used slave labour on a large scale, the labour for canals was provided by the superior land-holding sections on the canal route. The labour supplied by the superior land-holding sections would inevitably be comprised of tillers, artisans, service and menial labour castes, *sepi* as they were called in the Punjab, who were tied with agricultural production during the period under study. Among those whose labour produced material wealth, only a small section of the tillers of the soil might have gained marginal access to and benefit from the canals, the remainder either had no access to or use for the water their labour made available to others. Their labour generated increased surplus through stabilisation of agriculture, expansion of the area under cultivation and production of superior/cash crops which supported the state and the dominant sections of society.

III A

The field-specific mode of irrigation, the well,[67] known by different names in different parts of the Indian subcontinent, was extensively

[65] *Ibid.*, p. 214.

[66] *Ibid.*, p. 222, fn.3.

[67] The portion on the wells is substantially a reproduction of my article 'Rights to Sink and Repair Wells and Accruing Rights in Land and its

used and still continues to be used within the entire length and breadth of the subcontinent. While the prevalence of well-irrigation and the state policy of encouraging sinking of wells, the history of various technical devices, their dissemination, efficacy and costs and their impact on productivity have received attention from historians, the issue of rights and mechanisms of sinking and repairing wells largely remains a desideratum in historical writing on the subject. This may partly be owing to the fact that despite serious attempts by a few historians,[68] water rights have substantially remained outside the discourse of the history of the agrarian system in the subcontinent.[69] Also the fact that being a field-specific mode of irrigation, the wells may have been assumed to be part of the land system itself. The issue emerged clearly and powerfully when, during the course of my research on irrigation and social relations in the Punjab, I came across a document entitled the 'Right of Tenants to Sink Wells'.[70] The document defined the issue itself.

There were proprietary rights in wells. The ownership of the well belonged to the individual, family, community, authority depending

---

Produce', *Proceedings Indian History Congress*, 72nd Session, Patiala, December 2011, pp. 378–391.

[68] Notably B. D. Chattopadhyaya, *The Making of Early Medieval India*, New Delhi: Oxford University Press, 1997 and *Aspects of Rural Settlements and Rural Society in Early Medieval India*, Calcutta: Centre for Studies in Social Sciences, 1990 and David Ludden, particularly his article 'Patronage and Irrigation in Tamil Nadu: A Long-term View', *Indian Economic and Social History Review*, XVI: 3, pp. 347–65. James Heitzman, *Gifts of Power*; T. M. Srinivasan, *Irrigation and Water Supply: South India, 200 BC–1600 AD*, Madras: South Asia Books, 1991, is a monumental work on irrigation, but it is not much within the discourse of the agrarian system.

[69] Even when 'brain-storming' was done to go deeper into the agrarian system, it took into consideration three basic elements: 'land, labour and lord'. Water as a parallel component to land in the agrarian system escaped their notice. See Robert Eric Frykenberg, ed., *Land Control and Social Structure in Indian History*, pp. viii, xx. Water rights as a distinct entity have escaped even the notice of historians who have otherwise included irrigation as an important component in agrarian production and power structures and relations.

[70] 'Rights of Tenants To Sink Wells', in *Selections from the Records of the Office of the Financial Commissioner, Punjab*, 1874, pp. 236–240.

upon who had sunk the well. The well could also be owned by institutions such as temples. The state-sunk wells belonged to the state. The property in wells was inheritable and alienable; it could be leased, rented, mortgaged or bestowed. Each well had a title plate indicating its ownership. It is precisely owing to the title plate of the well that we have rich epigraphic evidence on the subject. Each well was known either by the name of the individual/caste/community/authority who had sunk it or by the purpose for which it was sunk. If the well was sunk outside of the state authority, permission was required. The tillers of the soil had no right to sink wells almost till the end of the nineteenth century.

There is varied and substantial evidence to support each one of the points made in the preceding paragraph. It comes from different sources from different parts of the subcontinent and is spread over nearly two millennia. There is a remarkable continuity and convergence in the entire range of evidence from different sources, places and time. The British settlement reports are replete with information on each one of the issues outlined in the preceding paragraph and several additional issues pertaining to wells and irrigation from wells. Once the issue gets defined through the British records, already known and existing evidence acquires a new meaning and yields information and throws light on the rights relating to wells. This is true of all genres of evidence: epigraphic, documentary and literary.

The ownership of the irrigation wells was not subsumed under the landholding. It was a property in its own right and was distinct from the land rights of a landholder. An irrigation well owned severally watered several landholdings and there were defined shares in its water. Even an individually-owned well serving only one estate was a property by itself, an entity distinct from the landholding.[71]

The British government recorded extensively and meticulously property in wells besides the landholdings during the course of making

[71] A few examples would suffice. G. Ousley and Capt. W. G. Davies, *Report on the Revised Settlement of the Shah Poor District in the Rawulpindi Division*, Lahore, 1866, pp. 111–12, no. 283 for shares in wells; S. S. Thorburn, *First Regular Settlement of the Bannu District,* pp. 95–97. See also J. H. Morris, *Report on the Revised Settlement of the Mooltan Division*, pp. 3, 5–6; 'Mozzuffurghur Settlement', in *Selections from the Public Correspondence of the Administration, for the Affairs of the Punjab*, vol. I, no. 2, Lahore, 1853, p. 21.

and revising land settlements in different parts of the Punjab, mostly in the second half of the nineteenth century.[72] These rights were recorded in view of the litigations that invariably followed any land settlement. In some districts it was noted that while litigation in severally-owned wells was frequent, it was not totally unknown in the individually-owned wells.[73]

In view of the information in the British records regarding property in wells and the fact that water rights attached to particular plots of land were often enumerated in the deed on the occasion when they changed hands by sale or transfer,[74] the information regarding wells in the documents from an earlier period of grant or sale of land assumes a new meaning. Thus, for instance, in a collection of 52 documents[75] from the Mughal and Sikh periods covering the time span 1695–1857 CE, there are seven documents that deal with the substantial question of grant, sale or lease of land[76] as distinct from grant of proceeds from land, their confirmation subsequently or issues relating to or emanating from such grants, sales or leases; one of these seven concerns culturable waste.[77] Of the six documents dealing with the developed land, five contain explicit information on the wells, information which is significant from the point of understanding property in wells as an entity distinct from the land in which they stood. Document number I, coming from Aurangzeb's reign, dated 1696 CE deals with a certain amount of land being given on *ijara*, land 'irrigated by well'.[78] The subject matter of document number II, dated April 1711 dealing with the sale of an entire village with 'valid proprietary and exclusive rights' records that this sale was 'with all its rights *and including, in full*,[79] fruit-bearing and non-fruit bearing trees and one *pucca* well of sweet water '.[80]

[72] See, for instance, 'Mozzuffurghur Settlement', *ibid.*, p. 21. Similarly all settlements recorded property in wells.

[73] See, for instance, J. H. Morris, *Report on the Revised Settlement of Goojaranwalah District in the Lahore Division*, 1860, p. 54, no. 82.

[74] Srinivasan, *Irrigation and Water Supply*, p. 163.

[75] B. N. Goswamy and J. S. Grewal, *The Mughal and Sikh Rulers and the Vaishnavas of Pindori*.

[76] These documents are numbers: I, II, IV, V, XXVII, XLIV, XLIX.

[77] Document no. XLIX.

[78] *Ibid.*, p. 78.

[79] Italics mine. *Ibid.*, p. 86.

[80] *Ibid.*

Document number IV of January 1738, registering the sale of land comprising the entire village, does not specifically refer to any well but qualifies it with 'each and every right'.[81] However, document number V of March 1738 dealing with the grant of a village 'with all its cultivated and uncultivated and, inhabited and unihabited land', adds that this was '*together with* all its fruit-bearing and non-fruit-bearing trees and one pucca well'.[82]

Similarly, document number XXVII of October 1789 records the grant of 24 *ghumaons* of land '*together with* a well, and an orchard'.[83] Document number XLIV of 1823 recording Ranjit Singh's *dharmarth* grant goes beyond all earlier documents in terms of information regarding the well in the document. The grant of the land is captioned in Gurumukhi as 'A well has been granted'.[84] The document records that 'some land together with a well, with a single Persian wheel, named after Amin Chand Thalwala' was granted in dharmath to the *mahant* of Pindori.[85] This alone validates all the points made in the preceding paragraphs. The fact that land grants/sales were being made 'together with' trees and wells is important in the context of the theme under discussion. There were distinct rights on trees and, therefore, the grant of trees was also specifically recorded.[86] For the same reason the grant of wells was being separately recorded and one well was recorded with its type and name. From another set of documents coming from Akbar's period from Vrindavan-Aritha there are similar references; there is one reference to a well by name[87] and another to a plot 'including well',[88] and yet others to

[81] *Ibid.*, p. 103.

[82] *Ibid.*, p. 111.

[83] *Ibid.*, p. 240.

[84] *Ibid.*, p. 312.

[85] *Ibid.*

[86] Tripta Wahi, 'Nature of Land Rights in the Pre-Colonial Punjab: A Study of the Tenancy Documents', *The Panjab Past and Present*, XXXVI, Part II, October 2005, pp. 2, 7–8.

[87] Tarapada Mukherjee and Irfan Habib, 'Land Rights in the Reign of Akbar: The Evidence of the Sale-deeds of Vrindaban and Aritha', in *Proceedings Indian History Congress*, Gorakhpur Session 1989–90, Document nos. 9–10, p. 247.

[88] *Ibid.*, p. 248.

trees.[89] In a document from the family archives of a *madad-i-maash* (revenue-free land grants given to men of learning and religious persons) assignee in the village Nandla in the pargana and suba Ajmer, the mention of both the lined (*Chah-i-Basta*) and unlined wells (*Chah-i-Kham*) would undoubtedly fall within the category of the wells within the Pindori documents.[90] From south India too, there are records of land being sold with wells. There is a reference to 3,000 *kuli* of land with five water levers being sold by the villagers to Ukkal during the period of Rajendracola Deva I.[91]

Our reading that land and well rights are treated as distinct entities in all these documents receives unambiguous support from the evidence coming from the reign of Firuz Shah Tughluq (1351–1388 CE). Two of Firuz Shah Tughluq's letters/orders in the *Insha-i-Mahru* deal with the subject of grant/its confirmation to two *khanqahs*. The first one reads 'the wells of the khanqah and of the neighbourhood of the Nahrwala town, which are related to the khanqah and for various reasons been under the control/right of Sayyid Muhammad were being given/confirmed' (the khanqah and the wells).[92] It was being ordered that all officials of Gujarat, from the *wali* to the *karkun* were to regard the khanqah and the wells as under the right/control of Saiyyid Muhammad.[93] A similar distinction is made in the next document regarding the khanqah of Kodia where rights over 'wells and land were granted'.[94]

There is also ample epigraphic evidence on the ownership of wells and rights in wells. This evidence exists precisely because each well carried a title plate and some of these title plates have been recovered, rescued and deciphered as inscriptions; many of these are from step-wells or otherwise large wells constructed by royalty. But there are a few inscriptions dealing with wells sunk by private individuals/communities, albeit those are also from fairly big wells. A close scrutiny

[89] *Ibid.*, Document no. 6, p. 246.

[90] B. R. Grover, 'The Extension of the Irrigation System...', p. 234, fn. 9.

[91] Srinivasan, *Irrigation and Water Supply*, p. 182. See also p. 163 for the sale of land with trees, well and tank.

[92] *Insha-i-Mahru, Tughluq Kalin Bharat,* vol. II, trans. S. A. A. Rizvi, Document no. 4, p. 376.

[93] *Ibid.*

[94] *Ibid.*, Document no. 5, p. 376.

of the entire *Epigraphica Indica* series is likely to yield more extensive evidence on the subject.

One of the earliest known inscriptions from a well is on a massive stone and it is ascribed to the first century BCE.[95] It is from Ghosundi in the Udaipur State. From 644 CE onwards there are numerous inscriptions from Rajasthan pertaining to different kinds of wells that have been used and noted by B. D. Chattopadhyaya. They are given in a tabulated form indicating chronology, sites and types of wells.[96] Similar kind of information is available for other parts and dynasties too.[97] A few examples may be cited; a Chalukya inscription of the *Vikrami Samvat* of 998 records the construction of a well[98] and some of the Pallava inscriptions give similar and more information about wells.[99] There is a remarkable continuity in the inscriptions from wells from the Sultanate and the Mughal periods. One of the earliest inscriptions from the Sultanate period is regarding the restoration and construction of a well by Qutlugh Khan in Bengal 'in the days of Iltutmish'.[100] There is an inscription recording 'the clearance and re-digging of a well' in Bayana in Bharatpur state from Balban's period under the governorship of Nusart Khan; this was done 'as scarcity of water was causing trouble to the people'.[101] The inscription was found in a well when it was being re-excavated.[102]

[95] *Annual Report on the Working of the Rajputana Museum, Ajmer, for the year ending 31st March 1939*, no..4 of the 'inscriptions inscribed', p. 4 (henceforth this series would be referred to as *ARWRM*).

[96] B. D. Chattopadhyaya, 'Irrigation in Early Medieval Rajasthan', *Journal of the Economic and Social History of the Orient*, XVI: 2–3, 1973, pp. 307–08.

[97] See, for instance, Srinivasan, *Irrigation and Water Supply*, pp. 101, 153, 163, 177–78, 183. In fact, the whole work is based primarily on inscriptions connected with irrigation.

[98] *Annual Report of the Sardar Museum and Sumer Public Library, Jodhpur for the year ending 30th September 1925*, p. 2. This inscription was fixed in a well.

[99] T. V. Mahalingam, *Inscriptions of the Pallavas*, New Delhi, Indian Council for Historical Research, 1988, no. 89, Tandantottam Plates of Nandi Varman II, p. 305, no.121, Velurpalaiyam Plates of Nandi Varman III, p. 379. See also Srinivasan, *Irrigation and Water Supply*, pp. 177–78.

[100] *Epigraphica Indo-Moslemica* (henceforth *EIM*), 1911–12, no. XXIII, ed. J. Horovitz, p. 25.

[101] *EIM*, 1937–38, G. Yazdani, ed, pp. 5–6.

[102] *Ibid.*, p. 5.

From Ghiasuddin Tughluq's reign there is a bilingual inscription from Petlad in Baroda recording the repair and completion of a well and the grant of 20 *kubhas* of land for its maintenance.[103] There is an inscription which is preserved in the Delhi Museum of Archaeology but which was originally found in a ruined well at Humayunpur near Hauz Khas. The inscription is engraved on a marble slab and records the building of a well named Chahi Khass during the time of Sikandar Shah Lodhi when Munnavar Beg was the governor of Delhi.[104] A Persian inscription of 1584 CE from Bhonsara in the 'Gwalior State' records the construction of a well by the order of the Mughal emperor Akbar.[105] There are several inscriptions recording the construction of big wells by the Kachchwahas, Sisodias and other Rajput chieftains during the Afghan–Mughal period; *Bhogidas ki Bavadi*[106] and the *baoli* at Joda Raya Simha are two such examples. The latter was built by Purohit Chakrapani and his sons during Salim Shah's time.[107] Among the numerous extant wells, the best preserved example is of a well built by the Muttaraiyar family, a subordinate of Dantivarman. The well is known as Marpidugu-Perunginaru well after the title of the Muttariyar family.[108]

While most of the wells were constructed by powers that be, there are several instances of individuals and communities constructing wells of different types. Thus, for instance, the Dabok inscription of 644 CE refers to the individual ownership of the *araghatta* fields. Also the Kekind inscription of 1143 CE refers to individual ownership of the arghatta fields.[109] Four centuries later, a baoli called Kalibaya near Khandela was constructed by one Agrawala Vania Prithviraja of Kolha and his son. The construction work was begun in *samvat* 1575 (1518 CE) during the reign of Sultan Ibrahim Lodhi and it was completed 17 years later (samvat 1592) during the reign of emperor

[103] *EIM*, 1917–18, no. 2, pp. 17–18.

[104] *EIM*, 1919–20, Horovitz, ed, pp. 8–9.

[105] *EIM*, 1937–38, pp. 22–26.

[106] *ARWRM Ajmer* for the year ending 31 March 1935, New Delhi 1936, 'Inscriptions copied', no. XIV. See also inscription no. XV.

[107] *ARWRM Ajmer*, year ending 31 March 1937, 'Inscriptions Copied, nos. XVII, XVIII, pp. 6–7.

[108] Srinivasan, *Irrigation and Water Supply*, p. 178.

[109] B. D. Chattopadhyaya, *The Making of Early Medieval India*, pp. 312, 315.

Humayun.[110] An inscription built into the wall of a baoli at Gyaspur in Pratapgarh records that the well was constructed by the banjaras in the samvat 1684 (1627 CE) in the time of Rawat Singha. Together with Nayak Gira of the banjara caste seven other family names are mentioned and one of them is from Agra.[111]

For individuals and communities to have sunk wells, permission to sink would have been acquired from the state. Presumably permission would have been acquired from the rulers whose names appear on the title plates of the wells. In the case of the Kalibaya baoli the names of two rulers appear, one of Ibrahim Lodhi during whose reign the construction work was begun and the other of Humayun in whose reign the work got completed. Presumably permission was acquired again from the new ruler with the overthrow of the dynasty in whose rule the construction had begun. Also because it was a huge work, renewal of permission would have been essential with the change of the ruling dynasty. Indeed, no well, particularly of such huge dimensions, could be constructed without the royal permission.

III B

Two of the earliest pieces of evidence on the right to sink wells come from the seventh and eighth centuries from the Pallava charters gifting land to *brahmadeyas*.[112] Besides the land, those who received land were also given the right to sink wells 'small and big' in the lands granted to them. The Anabil plates of Sundracola also give the donees the right to sink wells.[113] These rights were qualified by certain restrictions.[114] Obviously, the state controlled the construction of wells. In fact, there was double control over the construction of masonry wells. Besides controlling the right to sink wells, the state also controlled the right to the usage of burnt bricks. While the Tanjavir plate gives the right to

110 *ARWRM Ajmer*, year ending 31 March 1935. Inscriptions copied, no. X, pp. 4–5.

111 *Ibid.*, no. XII, pp. 5–6.

112 Mahalingam, *Inscriptions of the Pallavas*, Inscription nos. 89 and 121, pp. 305, 379. See also Srinivasan, *Irrigation and Water Supply*, pp. 177, 194.

113 Srinivasan, *Irrigation and Water Supply*, pp. 152–53, 177–78.

114 *Ibid.*, pp. 152–53.

use burnt tiles, Chitrur Plates give the donees the right to the usage of burnt bricks for houses and mansions.[115] In view of these controls, it is not surprising that most irrigational facilities, including the arghattas in Rajasthan emanated from the state. Obviously the right to sink wells rested with the state and it granted that right to others.

The state maintained a close watch and control over the wells through various administrative mechanisms. Thus, for instance, the formal sanction of the village governing bodies for sinking wells (particularly a large well) through the payment of a small cess called *ulliya-kuli* retained the control of the ruling power.[116] In south India there were inspectors of wells called *kupa-darskas*.[117] In Rajasthan there was a tax called *kosya*, presumably a tax on the *kosas* or leather buckets used for irrigation.[118] Also several cesses were levied on the araghatta fields; sometimes parts of the arghatta fields were earmarked for certain purposes.[119] Disputes in wells could receive attention from the rulers themselves; Maharaja Ranjit Singh's order to stop the sinking of a new well on another's land is on record.[120] At the village level, the *panchayats* could grant the right to sink wells.[121] While a cess was charged in south India for permission to sink wells, in the Punjab a *saropa* had to be given for sinking a well.[122] Even repair of wells was regulated; in some areas shares of repairs were defined,[123] in others applications were

[115] Mahalingam, *Inscriptions of the Pallavas*, p. 448.

[116] Srinivasan, *Irrigation and Water Supply*, p. 152.

[117] *Ibid.*, Glossary, p. 232.

[118] Dashratha Sharma, ed., *Rajasthan through the Ages*, Bikaner: Rajasthan State Archives, 1966, I, p. 330 (d).

[119] B. D. Chattopadhyaya, *The Making of Early Medieval India*, pp. 309, 311 and 314.

[120] J. S. Grewal and Indu Banga, *Civil and Military Affairs of Maharaja Ranjit Singh*, Document no. 58, p. 110.

[121] R. K. Saxena, *Peasant and the State: A Study of 18th Century Rajputana*, Jaipur: Panchsheel Prakashan, 1999, p. 291.

[122] Literally meaning from head to foot; a dress of honour; an installation fee paid for permission to sink a well. Indu Banga, *Agrarian System of the Sikhs*, New Delhi: Manohar, 1978, 'Glossary', p. 210; Charles A. Roe, *Customary Law of the Multan District regarding Inheritance, the Enjoyment of Property, Land Tenures and Alluvion and Diluvion*, Lahore: Victoria Press, 1883, pp. lxxxii-iii.

[123] V. Venkayya, 'Irrigation in Southern India in Ancient Times', *Archaeological Survey of India Annual Report 1903–04* (reprint), New Delhi, 2002, p. 210.

made to authority seeking permission to repair wells.[124] In south India there is ample evidence of village and supra-village bodies supervising irrigation facilities, including wells.[125] Firuz Shah Tughluq addressed himself to the entire hierarchy in Gujarat from the wali to the karkun regarding the grant of wells to the khanqah mentioned earlier. In the Multan region of the Punjab, and even elsewhere, the state was fully cognisant of the state of wells in the nineteenth century and so was it in different parts of Rajasthan. The issue will be discussed a little later.

It is apparent that there were well-organised structural mechanisms of controlling even field-specific mode of irrigation. They might have varied in form in different regions and at different times, but the systems of control were maintained by the state throughout the subcontinent. These systems of control could have been maintained directly or in partnership with different levels of authority/power/dominant groups as the situation would have required at any given point of time.

III C

The tillers of the soil had no right to sink wells. While it is understandable that they could not have had the means to construct masonry wells, what is remarkable is that they did not have the right to sink even a *kachcha*[126] well, which, in greater part of north India was 'no more than a hole dug in the ground to the depth of a few feet within a diameter of three to four feet'[127] with stake/wattle but invariably without it. A kachcha well is sunk down low enough to

[124] In the Chaj Doab in the Punjab, the British officer L. Bowring received numerous applications for permission to repair wells. *Foreign Miscellaneous H. M. Elliot Revenue etc. of the Punjab 1849*, no. 352, p. 53, NAI, New Delhi.

[125] One has just to read Srinivasan's book or even Venkayya's article to see how tightly controlled irrigation was.

[126] A *kachcha* well as distinct from a *pucca* well which has a wooden frame at the bottom, known as the *chak* in the Punjab, on which a bricked cylindrical wall is raised.

[127] Debate in the Legislative Council following the judgment of the Agra High Court in July 1867: *Home: Gazette of India Supplement*, July to December 1868, pp. 722–23.

ensure a good supply of water. The kachcha wells are not renewed or repaired, but have to be cleared out.[128] It is significant that the tillers of the soil had no right to sink even a kachcha well which only required their family labour and a small sum of money. Indeed, they were debarred from sinking wells.[129] The vital import of this denial of the right to the tiller of the soil and its *raison d'etre* would emerge in the following paragraphs. Suffice it to say here that the economic irrationality ascribed by Frykenberg to the villagers of lower standing for borrowing money not for improving productivity of land, but for supporting 'unproductive ceremonials and extravagances' is poignantly misplaced;[130] they did not possess the right to improve the productivity of the soil on their own. In fact, they were penalised for doing so as it would emerge from a judgment of the Agra High Court in 1867.

The right to sink wells was a highly prized right and it was closely guarded by those who possessed it. Sinking wells was one of the privileges of the superior land-holding class and the cultivators were debarred from that privilege.[131] It is repeatedly underlined in the customary law digests, settlement reports and the tenancy documents from the Punjab that the tenants did not have the right to sink wells. It was emphatically stated that the 'right of tenants to sink wells is recognised under no circumstances'.[132] In a report of the Law Committee of the *Anjuman-i-Punjab,* it was noted that in certain adverse circumstances the owner of land could be compelled by the need of securing tenants at any cost, including 'allowing them to exercise certain rights almost resembling proprietary rights over the land'. The *Anjuman* unanimously objected to the use of the expression 'almost resembling proprietary right' and

[128] E. B. Steedman, *Report on the Revised Settlement of the Jhang District,* 1874–1880, pp. 75–76.

[129] Tripta Wahi, 'Nature of Land Rights in the Pre-Colonial Punjab', p. 11.

[130] Frykenberg, *Land Control and Social Structure in Indian History,* Introduction, p. xiv.

[131] *Papers connected with the Question of Tenant Rights in the Punjab: Selections from the Records of the Government of the Punjab*, Lahore, 1869, pp. 94–95, 107, 276 (henceforth abbreviated as *Tenancy Documents*). See also Tripta Wahi, 'Nature of Land Rights', p. 11.

[132] T. Gordon Walker, *Punjab Customary Law*, Vol. V, *The Customary Law of the Ludhiana District*, Calcutta, 1886, p. 91, no. 182.

wished it to be substituted by an expressedly unambiguous statement that the tenants never acquired the right 'to sell trees or sink wells or transfer the land'.[133] It is apparent that the right to sink wells was a proprietary right. In one of the exceptions to the prevailing customs in the common holdings in the Punjab, sinking wells has been mentioned 'as exercise of one of the powers attending absolute ownership'.[134]

Control over rights to sink wells gave control over the tillers of the soil and the untouchables. Historic denial of rights in wells underlay untouchability; they were not allowed to draw water from the wells used by the upper castes. The untouchables were provided with wells much beyond the village boundary limits. Such wells as were specifically located for them were known as *parai-kulam* in south India.[135] Control over their labour followed control over the wells. As for the tillers of the soil, they were denied the right to sink even a kachcha well because rights in land accrued as a consequence of sinking wells. If cultivators' hold over their land became secure, it would weaken the control of the land-holding sections over them and thereby their control over the surplus.

The British government went in a tizzy following a judgment of the full bench of the Agra High Court in July 1867 whereby the court upheld the eviction of an occupancy tenant on the ground of his having sunk a kachcha well without the prior permission of the landlord.

The case which went through several stages of appeal was finally decided by the full bench of the Agra High Court on 20 July 1867.[136] The landlord's having ejected the tenant for having sunk a kachcha well without his prior approval was upheld by the full bench. It was argued that 'the act of digging a well or planting trees may not necessarily imply or assert a proprietary right in the land in which the well is dug or the trees are planted, yet by the general law of these provinces, a *ryot* even having a right of occupancy being prohibited from doing certain acts, such as planting of trees or digging of a well, without his landlord's consent, makes himself liable to ejectment'.[137] It was further

[133] *Tenancy Documents*, Anjuman's Comments, no. 5, p. 569. See also Tripta Wahi, 'Nature of Land Rights', p. 11.

[134] *Tenancy Documents*, no. 1, p. 132.

[135] Srinivasan, *Irrigation and Water Supply*, pp. 177–78, f.n. 111, p. 194.

[136] 'Right of Tenants to Sink Wells', p. 395.

[137] *Ibid.*

argued that 'the beneficial nature of an act is not a justification of it, if it be a breach of contract'. Further, 'A condition not expressly made between the parties to a contract, may nevertheless be attached to such contract by custom.' It concluded: '*The court must recognise the law as it is found to exist, so long as it shall not be superseded by positive law.*' As for the penalty, it concluded, '*the unwritten law of the country must be our guide*. Were we free to legislate upon the subject, it might seem to us equitable to look to the amount of injury actually caused to the landlord by the act complained of, and to grant him relief and compensation, whenever possible, otherwise than by the ejectment of the tenant. But it is not contended or proved that any other penalty than forfeiture of his holding for such a breach of contract is sanctioned by the law of these provinces'.[138] The case was decided with costs. The tenant by sinking the well had challenged the existing social relations of production.

The judgment created furore in the Legislative Council in its meeting held to discuss the Oudh Rent Act. That an act of improvement such as a 'mere hole dug in the ground' could lead to an occupancy tenant to lose his land was baffling for the members of the Legislative Council. They regarded the fact of digging of a kachcha well leading to the forfeiture of his occupancy rights as 'truly extraordinary'.[139] They failed to understand the implications of the issue for social relations.

IV

Indeed, primary proprietary rights existed in water. In some areas in the Punjab it was held that 'property exists essentially in water'.[140] The landholdings were known by its water, and the estates irrigated by wells, were defined by the nature of their wells.[141] In fact, superior rights in land accrued as a consequence of rights in water.[142]

It is evident from the judgment of the Agra High Court that the act of sinking a well was deemed an assertion of proprietary rights in

[138] *Ibid.*, pp. 397–98.

[139] *Home: Gazette of India Supplement*, July to December 1868, Oudh Rent Bill, Discussion in the Legislative Council on the judgment, pp. 720–31.

[140] *Tenancy Document*, p. 286.

[141] 'Mozzuffurghur Settlement', p. 15.

[142] Tripta Wahi, 'Nautre of Land Rights', p. 11.

land. In fact, superior rights in land and/or its produce accrued as a consequence of sinking and/or repairing wells. Although the tillers of the soil did not have the right to repair and sink wells, there is ample evidence to show that they did repair/sink wells. Presumably they did it with the permission of the *malik* or the ruler of the day. Under whatever circumstances the hereditary cultivator sank/repaired a well, the fact of his having done so gave him some rights in the land which were qualitatively different from and were superior to his right to hereditary cultivation. Not only were such cultivators allowed to remain in 'undisturbed enjoyment of their holdings', it was understood that by allowing a cultivator to sink a well at his own cost, the proprietors 'tacitly permitted him (cultivator) to get a title to cultivation even against themselves'.[143] Sinking a well could make him a joint proprietor of the land in which the well was sunk.[144] All kinds of exceptions to the existing rules and usages were permitted where a cultivator had sunk a well.[145]

With the sinking of a well the cultivator apparently became a proprietor.[146] It is noteworthy that those hereditary cultivators who had repaired/sunk a well invariably went beyond seeking the occupancy rights in the Punjab when the British made their first regular land settlement in the early 1850s; they sought *milkiyat* rights. Even when the entire question of occupancy rights in the Punjab was reopened by Edward Prinsep,[147] and he reversed three-fourths of the occupancy tenancies from the earlier settlements, 1,253 occupancy tenants were raised to the superior status of proprietorship.[148] This was done with mutual consent from both sides owing to the tenants having repaired/sunk wells. After repairing wells, even some untouchables acquired superior rights under Prinsep's settlement. Under Maharaja Ranjit Singh new tenures such as *chakdar, siledar, taraddakar, adhlapidar, kasurkhawar* developed arising out of sinking

143 *Ibid.*, pp. 9–13.

144 *Tenancy Documents*, see, for instance, p. 570; Tripta Wahi, 'Nature of Land Rights', pp. 11–15.

145 *Ibid.*, see, for instance, exception no. 1, p. 133, no. 14, p. 133, no. 32, p. 136.

146 *Ibid.* Byj Nath's reply, p. 96.

147 Tripta Wahi, 'Nature of Land Rights', pp. 5–6.

148 *Ibid.*

and repairing wells in the lands belonging to others. These tenures gave them rights in land and/or its produce.[149] The nomenclature of these tenures is derived from the well itself; *chak* is the wooden frame at the bottom of the well and the *siledar* is derived from the word for brickwork of a *pucca* well.[150] There were nuanced differences between and within these tenures depending upon the circumstances and conditions under which a particular well was sunk, but all these tenures were inheritable and alienable.[151] It would be worth examining whether such tenures developed in other regions too where the state towards the end of the eighteenth century was attempting to get the trading communities, as in the Punjab, to invest money in repairing and sinking wells.[152]

It is quite remarkable that even after the landlords had acquired indefeasible proprietary rights in land under the British, they were apprehensive that by sinking wells cultivators would acquire superior rights in land. They put up stiff resistance to the attempts by the British government to give hereditary tenants right to sink wells. In the Punjab, they were not agreeable to tenants being given the right to sink wells unless they 'put the proprietors' name on the title-brick or sign a written agreement with the proprietor to prevent the possibility of the latter ever losing his right of property.[153] As late as 1911 in the United Provinces they were demolishing the masonry wells constructed by the tenants[154] notwithstanding the fact that the tenants had been granted that right by the British government.

[149] J. H. Morris, *Report on the Revised Settlement of the Mooltan District*, Lahore, 1856–7, pp. 6–8, 28, Appendix H. no. 7; A. R. Roe, *Report on the Revised Settlement of the Multan District of the Punjab*, pp. 40–41, Henry Monckton, *Report on the Revised Settlement of the Jhung District*, p. 4, nos. 27–28; see also Indu Banga, 'Ecology and Land Rights in the Punjab', pp. 64–65.

[150] *Ibid.*

[151] Charles A. Roe, *Customary Law of the Multan*, pp. lxxxii–ix; see also, Indu Banga, *ibid.*, pp. 64–65.

[152] See Dilbagh Singh, *The State, Landlords and Peasants: Rajasthan in the 18th Century*, New Delhi: Manohar, 1990, p. 53; R. K. Saxena, *Peasant and the State*, p. 11.

[153] *Tenancy Documents*, pp. 72, 81, 95.

[154] Irfan Habib, *A People's History of India 28: Indian Economy, 1858–1914*, New Delhi: Tulika Books, 2006, p. 51.

V

Tight control over the right to sink and repair wells was an essential component of the social relations defining access to means of production with the aim of controlling and meticulously regulating the release of the productive forces.[155] This introduced a fundamental contradiction between the interests of the ruling powers to enhance their resources and simultaneously uphold the social relations which controlled and regulated the release of productive forces. The nature of the state in the last resort would be defined by its policy on water rights.

With the centrality of irrigation and of water rights in agrarian production and generation of surplus, it is not surprising that water was an essential component in the system of revenue assessment and classification of cultivable land under the Mughals, a part of which the Mughals had inherited from their predecessors. Akbar's classification of cultivable land into, *polaj, parauti, chachar,* and *banjar* was based upon continuity or intermittence in cultivation; the polaj land was under continuous cultivation with successive crops and was never allowed to remain fallow while parauti was left out of cultivation for sometime so that it could recover its strength.[156] The chachar land was that land that had remained fallow for three or four years while the banjar had remained uncultivated for five or more years.[157] Water/irrigation was the basic criterion underlying this classification based upon the periodicity of cultivation. For the chachar and banjar categories excessive rains and inundation are explicitly given as causes for the land falling out of cultivation both in the river belts and the submontane regions. It is apparent that these two categories of land are dealing with inundation (*sailabi*) and rain-based (*barani*) cultivation; both sailabi and barani were unirrigated lands. As distinct from the chachar and the banjar categories, the polaj and parauti were under continuous/nearly continuous cultivation and continuity and stability in agriculture was in direct proportion to human intervention in water. It may be safely

155 Tripta Wahi, 'Land Rights in the Pre-Colonial Punjab', pp. 7–8, 13.

156 *Ain* XI, 'Land and its Classification and the Proportionate Dues of Sovereignty', in *Ain-i-Akbari*, II, p. 68.

157 *Ibid*. See also, B. R. Grover, 'Classification of Agrarian Land Under Cultivation', pp. 241–42.

inferred that irrigation underlay basic classification of cultivated lands for calculating 'proportionate dues of sovereignty' under Akbar. Within the irrigated territories, that is, the polaj and the parauti, there were soil-related subcategories.[158] The Mughals had inherited these categories from their predecessors, the Sur dynasty.

Indeed, water as an independent and parallel component to land in calculating 'the dues of sovereignty' has been explicitly enunciated in the *Ain VII* on *Khazandar*. We may quote Abu'l Fazl for that:

> And because the conditions of the royal state and prerogative vary in different countries, and soils are diverse in character, some producing abundantly with little labour, and others the reverse, and as inequalities exist also through the remoteness or vicinity of water and cultivated tracts, the administration of each state must take these circumstances into consideration and fix its demands accordingly.[159]

That water and irrigation formed an integral part of the system of revenue assessment is apparent, but how they were factored in assessment awaits research. It is a distinct possibility that administrative/fiscal units/rates could have revolved around specificities of water/irrigation. It may not be out of place here to mention that the two divisions of *Birun-i-Panjnad* (outside of the rivers) in the suba Lahore[160] make sense only in relation to water-related factors. Separated by a large physical distance, with one located to the west of the river Indus and the other to the east of the Sutlej, these two divisions could not have formed one administrative unit;[161] their nomenclature suggests that they represented one category in relation to running surface waters of the Indus system of rivers of the suba of Lahore. Under Ranjit Singh the well was taken as a unit of revenue-yielding capacity of the land that it cultivated. There are numerous documents extant from Ranjit Singh's period which refer to grants of wells of varying values as *dharmarth* or for other purposes.[162]

[158] *Ibid.*

[159] *Ibid.*, *Ain VII* on *Khazandar,* pp. 58–59.

[160] Irfan Habib, *An Atlas of the Mughal Empire*, New Delhi: Oxford University Press, 1982; Punjab, Political 1595, Sheet no. 4A.

[161] *Ibid.*, Notes on Sheet no. 4A, p. 8.

[162] J. S. Grewal and Indu Banga, *Civil and Military Affairs of Maharaja Ranjit Singh*, See, for instance, Document nos. 123, 124, 144, 183, 197, 199, 238,

Indeed, irrigation, through its implications for agrarian production and generation of surplus, had wide and multi-dimensional ramifications for the socio-political structures, politics and polity of any region/state. Even a brief study of irrigation on the river Ghaggar in the first half of the nineteenth century reveals that there were disputed claims by different villages, zamindars and chiefs over waters of different streams and torrents, over right to construct *bandhs* or embankments at different points on the river and/or to make cuts/channels at different points.[163] There were rights even over the superfluous water and winter freshets. These rights were both defined and contested and they generated tremendous social tensions between villages and chiefs and invariably led to violence and even loss of life. Disputed claims and their resolutions reveal arena of social tensions and conflict and the nature and role of different institutions and levels of authority within that. Dynamics of larger political situations had direct bearing on these contestations and their resolutions; fluid political situations accentuated contestations leading to emergence of newer rights, while larger and more stable political entities invariably intervened to determine, regulate and also change these rights. History of each region and sub-region awaits study with water as an essential component in it.

At the higher political level, the politics of revenue assignments to the ruling elite would be influenced by the level of development of irrigational facilities or their absence; assignees with less developed irrigational facilities are known to have paid a great deal of attention to developing irrigation works in their assignments. In fact, the state itself would have had a great stake in creating and improving irrigation facilities.

It were not only the Delhi-centric big and powerful states that devoted attention to canal-making; this was the accepted norm everywhere within the subcontinent. Several centuries prior to Akbar

---

244, 262, 274, 287, 372. All have varying values or contexts. Document no. 287, besides giving the total value, mentions the value of the well for *kharif* and rabi crops separately.

[163] See Tripta Wahi, 'Socio-Political Structures in Interaction with a Seasonal River: A Case Study of the Ghaggar in the Mid-Nineteenth Century', in *Proceedings of the Punjab History Conference*, 39th Session, 2007, pp. 291–95. See also J. D. Cunningham, 'Report of the Irrigation of the Gugur and the Sursootee'.

and from a different part of the subcontinent, is a Tamil poet's advice to a Pandya king on the subject of canal-making:

> Oh! great king, if you crave wealth in the next world
> and yearn to vanquish other kings who protect this world
> and thus to become the greatest among them
> hearing songs of praise to your glory,
> listen to me to learn what deeds guarantee these rewards.
> Those who give food give life to living beings
> who cannot live without water.
> Food is first for all living things, made of food,
> and because food is but soil and water mingled together,
> those who bring water into fields
> create living beings and life in this world.
> Even kings with vast domains strive in vain, when their land is dry
> and fields sown with seeds look only to the sky for rain.
> So Pandya king who makes dreadful war, do not mistake my words:
> quickly expand watery places that are built to bring streams to your land!
>
> *For those who control water reap rewards*
> *and those who fail cannot endure.*[164]

Akbar's preamble to the *sanad* quoted above is reminiscent of the Tamil poet's advice to a Pandya king given several centuries earlier. It is remarkable how the two concur in their concerns; both show people-centric concerns, but in the last resort the interests of the rulers/state underpin their views on bringing 'streams' to the fields. While one views 'control over water' essential for survival of political powers, the other eyes 'hundred fold' increase in agrarian production and consequent increase in the imperial revenues. Indeed, rulers have always been deeply involved in large-scale irrigational activities and structures, systems and rules concerning their use have been in place for centuries before Akbar. As long back as Kautiliya's time, the Head of the Department of the Settlement of the Countryside was meant to 'cause irrigation works to be built with natural water sources or with water to be brought in

[164] *Purananuru*, no.18, lines 13–30, quoted in David Ludden, *An Agrarian History of South Asia, The New Cambridge History of India*, IV, Cambridge: Cambridge University Press, 1999, p. 78.

from elsewhere'.[165] There were rules governing any one walking out of the joint building of an irrigation work and[166] there were different water-rates for different kinds of irrigation facilities.[167] Not only was the administration supposed to keep record of water-works, but their regular inspection was also mandated.[168] Indeed, there are specific regulations given in the *Arthasastra* regarding control over ownership of irrigation works, usage of water, damage to the works or to the fields through which they pass, nature of revenue obligations for those making irrigational works, rules concerning renting, mortgaging and sale of these facilities and punishment for violation of those rules.[169] If from the *Arthasastra* we get an insight into the systems, structures and rules concerning irrigation, from the *Kaushiksutra* we get a glimpse of the rituals and ceremonies connected with the opening of a canal. When 'a canal was opened gold plate was laid at the mouth of the channel… and a frog was tied...'[170] Indeed, information is forthcoming from all genres of sources, regions and times. It is squarely staring us in the face to be picked up, studied and integrated within the discourse of the history of the Indian subcontinent.

The centrality of irrigation both for the state and the society is writ large on the face of Indian history, but what is not so apparent is the nature of water rights. In fact, the nature of rights of access to water is hidden from the view; there was inequitable access to water in the pre-colonial times and this was an essential component of social relations.

[165] *The Kautiliya Arthasastra,* Part II, trans. R. P. Kangle, Bombay: Motilal Banarsidass Publishers, 1963, p. 64, 2.1.20, p. 209, 2.34.8.

[166] *Ibid.*, p. 64, 2.1.20.

[167] *Ibid.*, p. 173, 2.24.18.

[168] *Ibid.*, p. 210, 2.35.3, p. 217, 2.36.45.

[169] *Ibid.*, pp. 249, 253, 255–56.

[170] Quoted, K. L. Rao, *India's Water Health: Its Assessment, Uses and Projections*, Hyderabad: Orient BlackSwan, 1975, p. 114.

Changing River Courses in North India*

Calamities, Bounties, Strategies—Sixteenth to Early Nineteenth Centuries

Meena Bhargava**

Understanding disasters—theorising and debating them—has actively engaged not only anthropologists, cultural geographers, sociologists and political scientists but also historians. Earlier, historians had neglected natural catastrophes and disasters as historical events, dismissing them as mere 'accidental facts'. They had argued that man was the sole 'actor' of history.[1] If disaster has not been studied in detail or in depth, one of the problems is that disaster is considered to be an event and not a process.[2] Yet another reason could be, as Stephen Kroll-Smith has argued, that 'disasters are good to think' but they are also difficult to think.[3] Scholars have found it problematic to reach a common understanding or a consensus on the definition of disaster. Sociologists define disaster as organisational behaviour, geographers call it hazard, and political scientists speak of risk assessment policies and practices.[4] Disasters, therefore, cannot be categorised neatly, either in type or scale.

* Meena Bharagava, 'Changing River Courses in North India: Calamities, Bounties Strategies – Sixteenth to Early Nineteenth Centuries' in *The Medieval History Journal*, Special Issue: Coping with Natural Disasters in Pre-Industrial Societies, volume 10, issue numbers 1–2, January–December 2007, pp. 183–208.

** **Acknowledgement:** To the memory of John F. Richards, who introduced me to environmental history. It is unfortunate that he did not live to see this article in print but I am happy that he could read it online.

1 Michael Kempe and Christian Rohr, 'Introduction', *Environment and History*, special issue 9: 2, 2003, pp. 123–26.

2 Anthony Oliver-Smith, 'Theorizing Disasters: Nature, Power and Culture', in Susanna M. Hoffman and Anthony Oliver-Smith, eds, *Catastrophe and Culture: The Anthropology of Disaster*, Oxford: James Currey, 2002, p. 23.

3 *Ibid.*

4 *Ibid.*, p. 24.

Recent analyses of disasters explain why they should not be separated from everyday life and indicate how the risks involved in disasters should be connected with the vulnerability created for many people through their normal existence. They therefore attempt to understand the links between the risks that people face and the reasons for their vulnerability to hazards.[5] Disasters are multidimensional, affecting most aspects of community life because they are both physical and social processes, in which a geophysical or biological event is evidently implicated in some form or the other in causing disaster.[6] Primarily, disaster is a social phenomenon and thus identifiable in social terms.[7]

Disasters have now become central to international comparative studies. In fact, most debates on human–environment relations and issues of development and sustainability hinge around disaster studies. The terms natural hazards and natural disasters emphasise the role of geophysical processes or an extreme geophysical event which act as the triggers of disaster. Disasters are thus the result of several social and economic factors and are often complex in their origins.[8] Entrenched deeply in environmental and human systems, disasters are an indicator of a society's failure to adaptation and sustainability.[9] Hewitt, however, has rejected the oft-held theory that disaster means the collapse of the productive potential of a social order. Shifting the focus from the disaster event, he emphasises societal and man–environment relations and defines natural disasters as the normal order of things rather than as an accidental geophysical feature.[10]

In analysing disaster and catastrophe, anthropologists have studied the construction of cultural meanings and world views. Faced with

[5] Piers Blaikie, Terry Canon, Ian Davis and Ben Wisner, *At Risk: Natural Hazards, People's Vulnerability and Disasters*, London and New York: Routledge, 1994.

[6] *Ibid*., pp. 4–5.

[7] E. L. Quarantelli and Russell R. Dynes, 'Response to Social Crisis and Disaster', *Annual Review of Sociology* 3, 1977, pp. 23–49.

[8] Keith Smith and Roy Ward, *Floods, Physical Processes and Human Impacts*, Chichester: John Wiley & Sons, 1998, p. 19.

[9] Anthony Oliver-Smith, 'Anthropological Research on Hazards and Disasters', *Annual Review of Anthropology*, 25, 1996, pp. 303–28.

[10] K. Hewitt, ed., *Interpretations of Calamity*, Boston: Allen & Unwin, 1983, pp. 24–27; also Oliver-Smith, 'Theorizing Disasters', p. 27.

change, loss and destruction, the disaster-stricken have raised existential questions which reflect the moral and ethical values of belief systems and include concepts of 'social and cosmic justice, sin and retribution, causality, the relationship of the secular to the sacred and the existence and nature of the divine'.[11] Natural disasters were interpreted as a sign of divine punishment in many ancient civilisations, both in the East and the West. It was believed that earthquakes, fires, landslides, floods and pestilence were caused by God's rage against the sinful lives of the people, the incompetence of the sovereign and the general moral decline in society. A disaster was considered a challenge from God to test the human capacity to manage it through truthfulness and righteousness.[12] The *Rigveda* and *Atharvaveda*, the two ancient Indian scriptures, recommend expiation by performance of religious rites such as worshipping of the Gods, chanting of sacred mantras and animal sacrifices.[13]

Disasters are natural calamities often considered to be extreme material events, while hazards refer to the potential risk that could lead to catastrophe. India is considered to be one of the most disaster-prone regions with about 85 per cent of the country prone to some kind of disaster, able to cause destruction and bring havoc to the physical environment and the resources of its society. For instance, the flood-prone area in India is estimated to be one-eighth (40 million ha) of its entire geographical area; one-fifth of the flood-prone area is subjected to floods every year.[14] But floods and inundations also contribute to land building by depositing silt and therefore to economic

[11] Oliver-Smith, 'Anthropological Research on Hazards and Disasters', p. 308.

[12] Christian Rohr, 'Man and Natural Disaster in the Late Middle Ages: The Earthquake in Carinthia and Northern Italy on 25 January 1348 and its Perception', *Environment and History*, Special Issue 9: 2, 2003, pp. 127–49; also C. M. Agrawal, *Great Natural Calamities*, New Delhi: Indian Publishers & Distributors, 2000.

[13] If there was no rainfall, Indra, the God of Rain had to be worshipped and in case the river changed its course or inundated its bank or was flooded, worship of that particular river had to be performed; see Agrawal, *Great Natural Calamities*, pp. 3–4.

[14] R. B. Singh, 'Disaster Monitoring and Mitigation', in R. B. Singh, ed., *Disasters, Environment and Development*, New Delhi and Calcutta: Oxford &

benefits and material well-being. To understand the several aspects and multidimensionality of natural disasters, we will in this chapter, focus on the nature and impact of the changing course of rivers in north India, in particular river Ganga (Ganges).

Water courses have played a significant role in determining human settlements and human activity. This perhaps explains the abundance of archaeological sites along river margins and the banks of streams and lakes. However, before ascribing any importance to the river banks or water channels from the point of view of human ecology or archaeology, we need to understand the basic geomorphologic processes that mould these micro-environments.[15] This has entailed, as Braudel has argued, 'a necessary reduction of any social reality to the place in which it occurs'.[16] If the social reality is perceived in this manner, geographic structures provide a context for different movements of time (especially the *longue durée*) and through them all categories of social life come to be mediated.[17] Such a methodology provides space 'for an increasingly geographical conception of mankind', as argued by Vidal de la Blache in 1903.

Archaeological evidence has rendered Blache's perception into a reality. It has emphasised the value of geography to history, especially in understanding the sites of settlements and demographic movements. These findings exemplify that geographical regions cannot be

---

IBH Publishing Co. Pvt. Ltd., 1996, pp. 43, 56; also 'Introduction' in Harsh Gupta, *Disaster Management*, Hyderabad: Universities Press, 2003; also Piers Blaikie, 'Floods', in Piers Blaikie, Terry Canon, Ian Davis and Ben Wisner, *At Risk*, pp. 124,131–32.

[15] Compare William A. Turnbaugh, 'Floods and Archaeology', *American Antiquity* 43: 4, 1978, pp. 593–607.

[16] Fernand Braudel, 'History and Social Sciences: The *Longue Duree*', in Maurice Aymard and Harbans Mukhia, ed., *French Studies in History, The Inheritance*, vol. I, Hyderabad: Orient BlackSwan, 1988., p. 99. For more details on *longue durée* and geographic structures also see Fernand Braudel, *Mediterranean World in the Age of Philip II*, London: Collins, 1972; and Emanuel Le Roy Ladurie, *The Peasants of Languedoc*, Illinois: University of Illinois Press, 1974.

[17] Braudel, 'History and Social Sciences', p. 99; also see Meena Bhargava, *State, Society and Ecology: Gorakhpur in Transition, 1750–1830*, Revised Edition, New Delhi: Primus Books, 2014 (Manohar, 1999), p. 10.

understood as permanent pivots to historical activities. Changes in landscapes occur whether owing to geology, geomorphology or human activity.[18] While there is a need to study the changes in landscapes, it is even more important to understand the impact of changing landscapes on history and one of the most effective ways of doing so is to examine the changes in river courses and their effects.

Rivers and Geomorphology

In any discussion of geomorphology and ecology, our focus is on nature and its bounty, and that despite its hazards and destructive capacity, it does not completely exhaust the earth.[19] While analysing geographic structures in any country, rivers come first, and so have a primary significance in any discussion of ecology. A river represents a stage of water circulation and is an indispensable module of the organic and inorganic world.[20] It may be observed that a river system includes all such streams which rise within a given hydrographic basin whose waters pass by one channel into the ocean reservoir. Affluents diverted from the main stream by delta accumulations or diverted into the sea by new courses are also considered a part of the general river system.

Rivers are thus natural drainage channels that carry away part of the precipitation that falls on the earth's surface. The size and shape of the river channels change and adjust constantly to the variable of precipitation and run-off. The extent and the type of adjustment are generally determined by the variation in precipitation levels and the physical characteristics of the terrain. Most rivers develop a main channel, sufficient to contain the volume of flow that occurs annually. Adjoining the main channel, there is usually a well-defined flood plain which is inundated at irregular intervals, depending on variable

[18] See Romila Thapar, *Early India: From the Origins to AD 1300*, London The Penguin Press, 2002, p. xxi.

[19] Bhargava, *State, Society and Ecology*, p. 10; also see Sugata Bose, *Peasant Labour and Colonial Capital: Rural Bengal since 1770*, Cambridge: Cambridge University Press, 1993, p. 14; and Christopher A. Bayly, *Rulers, Townsmen and Bazaars: North Indian Society in the Age of British Expansion, 1770–1870*, Cambridge: Cambridge University Press, 1983, p. 74.

[20] Bose, *Peasant Labour and Colonial Capital*, p. 9.

precipitation. These periodic inundations are natural phenomena that go back to long before when human intervention began to alter the environment. Inundations have left their impact and legacy through the ages, both on the land itself and in the history of man's use of the land.[21]

Flood hazards vary in intensity and severity and can occur in almost any river valley at any time due to climatic variability and extreme vulnerability. They are mostly unpredictable, either on a short-range or a long-range basis and are therefore considered a constant hazard.[22] The damages wrought by floods can often be attributed to socio-economic factors, particularly when despite warning, human settlements and anthropogenic activities cannot be moved away from the flood-prone areas.[23] The data suggests that the establishment of settlements on and in the neighbourhood of rivers and streams was not random. Settlement patterns indicate a human tendency to seek proximity to stream and river activity; rivers evidently being an essential source of water supply. For several years in the past, travel was virtually unthinkable except by water or on trails, roads and rail networks adjacent to the rivers.[24] These reasons remain compelling even today for the continuing use of valleys and flood plains, despite the damage and disaster caused by floods.

To understand the corrosive power of rivers and the relation of river action to geographic forms, we will attempt a case study of one of the major river systems in South Asia—the river Ganga and its tributaries and distributaries in north India. Rivers have histories like anything

[21] Albert P. Brigham, 'Rivers and Evolution of Geographic Forms', *Journal of the American Geographical Society of New York* 24, pp. 23–43.

[22] Samuel D. Sturgis Jr., 'Floods', *Annals of the American Academy of Political and Social Science* 309: 1, January 1957, pp. 15–22.

[23] William H. Ballou, 'Floods: Their History and Relations', *The American Naturalist* 19: 12, 1885, pp. 1159–62; also Paul B. Sears, 'Natural and Cultural Aspects of Floods', *Science* 125: 3252, 1957, pp. 806–07; also Blaikie, 'Floods', p. 124.

[24] R. B. Singh, 'Flood Prevention and Mitigation in Disaster Prone Regions of India: Case Study', in Frank Toensmann and Manfred Koch, eds, *River Flood Defence*, vol. 2, no. 9, Reports of Hydraulic Engineering, Kassel, 2000, p. G-157; also Turnbaugh, 'Floods and Archaeology', p. 604; and Sturgis Jr., 'Floods', pp. 17–18.

else and the river Ganga, like several other rivers of the world, has not remained static over the centuries or over decades. It has, in fact, played an active role in the lives of its people and earned a place as both a dynamic but also destructive personality in Indian folktales and songs.[25] The vulnerability of the region from where the river Ganga flows has increased owing to geomorphic, economic and social reasons.

Historical Sources

To understand the geomorphology of the river Ganga, I have largely relied on geological and archaeological interpretations. However, to analyse aspects of disaster or the positive and beneficial effects of the changing course of the river, Persian sources, memoirs, diaries, traveller's accounts and English archival records have been used. All historical sources have their own biases and limitations. For instance, although the Persian sources are rich in their description of the ecology and the environment—of the changing patterns, floods and disaster, as well as the material benefits of the river systems—they are, for the most part, state-oriented, the military and political concerns of the Mughal Empire are dominant and there is not much reflection on the views and perceptions of the people or the impact of disaster on them. Notwithstanding such bias, different sources have been used as a complement to each other to construct a view—disaster and dynamism—of the rivers in north India, in particular the Ganga.

Geomorphology of River Ganga

In their study of the shifting course of rivers, geologists Peter Clift of the University of Aberdeen and Jerzy Blusztain of Woods Hole Oceanographic Institution have argued that the Sutlej, Ravi, Chenab and Jhelum rivers, which today flow westward to join the Indus and

[25] Compare with river Terek in North Caucasus, its significance for the Cossacks and the extent to which they personalised it, giving it an important place in its folklore. See, Thomas M. Barrett, 'The Land is Spoiled by Water: Cossack Colonisation in the North Caucasus', *Environment and History* 5: 1, 1999, pp. 27–52.

drain into the Arabian sea, once flowed eastward into the Ganga and drained into the Bay of Bengal. The rivers changed their course some five million years ago, probably owing to the ongoing collision of the Indian and Eurasian tectonic plates that bulldozed the terrain, tilting it westward or lifting up river-diverting mountains. These scientists observe that the rivers which were once tributaries of the Ganga were 'captured' by the Indus. In their opinion, it is for the first time that 'such a major sediment capture event has been dated'. It, therefore, provides a potential to use ancient sediments to reconstruct what the mountains may have looked like in the past.[26]

The river Ganga flows through three countries—India, Nepal and Bangladesh. In India, it flows across the plains of north India (called the Gangetic plains) and forms the major river basin of that region. There is a general consensus amongst geologists that the Ganga plain is part of the Indo-Gangetic foreland basin system and was formed following the collision of the Indian and Asian plates. It has been suggested that the geomorphic features of the Ganga plain are a consequence of climatic change, base-level change and tectonic activity during late Quaternary.[27] The river channels of the Ganga plain were formed under different hydrological conditions and represent distinct stages of evolution and rejuvenation of channels.[28]

The Ganga plain thus shows a number of geomorphic surfaces of regional extent and genetic significance. It is geomorphologically diverse and drained by numerous south-easterly flowing Himalayan and plains-fed rivers.[29] The Ganga—the trunk river of the Ganges plain—is joined by large Himalayan rivers from the north except river Yamuna

[26] Peter Clift and Jerzy Blusztain, 'Changing the Course of Rivers and History: Rivers in Pakistan's Punjab Region Switched Directions', *Nature* 15 December 2005 in *Oceanus* (The Online Magazine of Research from Woods Hole Oceanographic Institution), p. 1.

[27] I. B. Singh, M. Jaiswal, A. K. Singhvi and B. K. Singh, 'Rapid Subsidence of Western Ganga Plain during Late Pleistocene: Evidence from Optical Dating of Subsurface Samples', *Current Science* 84: 3, 2003, pp. 451–54.

[28] I. B. Singh, 'Late Quaternary History of the Ganga Plain', *Journal of Geological Society of India* 64: 4, 2004, p. 431–54.

[29] U. K. Shukla and D. S. Bora, 'Geomorphology and Sedimentology of Piedmont Zone, Ganga Plain, India', *Current Science* 84: 8, 25 April 2003, pp. 1034–40.

(one of its tributaries), which meets the Ganga from the south. There are also a number of streams originating in the alluvium, joining the major streams. The rivers of the Ganga thus show a wide range of sizes and channel patterns.[30] To the north of the Ganga are its other large tributaries—Gomti, Ghaghara, Kosi, Gandak, Son, Ramganga, Mahananda, and further east its dual streams—Bhagirathi and Padma. Beyond Mahananda, the Ganga enters its own delta, formed by its distributaries and then merges into the combined delta of the rivers Ganga, Brahmaputra and Meghna.

The Ganga mega watershed is considered to be a sacred landscape, traversed by the Ganga river system. The Ganga river system in India, originating at Gaumukh in the higher regions of Garhwal Himalaya, passing through the northern plains of the states of Uttar Pradesh, Bihar and West Bengal and draining into the Bay of Bengal in the east, represents one of the best examples linking highland/lowland interactions.[31]

In his account of river Ganga, James Rennell wrote in 1781 that the bed of the Ganga was very unequal in point of width. From its first arrival in the plains at Hardwar (Uttar Pradesh) to the conflux of the Yamuna, its bed was generally from a mile to a mile and a quarter wide; and compared with the latter part of its course, tolerably straight. From here downstream, its course became more winding and its bed wider, till it alternately received the waters of the Ghaghara, Son, Gandak and many other streams.[32] The windings of the Ganga in the plains were owing to the looseness of the soil there, testified to by their perpetual changes. The courses of the rivers became more winding as they approached the sea. But this happened only if the rivers took the latter part of their course through a sandy soil. In the case of the Ganga and other rivers, subject to considerable and frequent variations, the best way to assess the vicinity of the sea was the low-level of river banks and the increasing muddiness of the shallows in the river beds.[33]

[30] I. B. Singh, 'Quaternary Palaeoenvironments of the Ganga Plain and Anthropogenic Activity', *Man and Environment* xxx: 1, 2005, pp. 1–35.

[31] Gupta, 'The Concept of Sacred Landscape', in Gupta, *Disaster Management*, p. 5.

[32] James Rennell, 'An Account of the Ganges and Burrampooter Rivers', *Philosophical Transactions of the Royal Society of London* 71, 1781, pp. 87–114.

[33] *Ibid.*, p. 98, 98n; Compare with the lower Po plain in Northern Italy in the sixteenth century, see Emanuela Guidoboni, 'Human Factors, Extreme

Over the course of millennia, the Ganga (which originates in the central Himalayas and flows 2,500 km to the Bay of Bengal) and its tributaries have constituted one of the largest flood plains in the world, accumulating sediments from the erosion of mountainous areas. The lower Ganga basin contains an active as well as a moribund delta, both of which have been affected by enormous flood flows in the Ganga, continuous tidal pressures from the seas and occasional severe cyclonic storms which have the potential of disrupting the environmental systems.[34] In the middle Gangetic plain, the general slope of the Ganga is from north-west to south-east but from Allahabad (Uttar Pradesh) onwards, the river enters flat plains which contain many *khadar* or flood plain zones.[35] The diversion of the river routes and the abandonment of their beds in this region have led to the formation of lakes (called *chars* in the local dialect of north Bihar) in the plains of eastern Uttar Pradesh and north Bihar.[36] The khadar region of river Ganga is partially a boon to the cultivators. Despite the virtually annual devastation caused by floods, this zone is extremely fertile and therefore with great potential for agriculture.[37] In the khadar of the upper Ganga (in Muzaffarnagar, Meerut and Bijnor districts of Uttar Pradesh), the course of the Ganga is subject to constant change. It has been observed that the river bed has shifted 3 km east in the 50 years between 1914 and 1964. Around

---

Events and Floods in the Lower Po Plain (Northern Italy) in the 16th Century', *Environment and History* 4: 3, 1998, pp. 279–308.

[34] Haroun Rashid and Babar Kabir, 'Water Resources and Population Pressures in the Ganges River Basin', in Alex de Sherbinin and Victoria Dompka, eds, *Water and Population Dynamics: Case Studies and Policy Implications*, accessed from www.aaas.org on 3 May 2006.

[35] R. S. Sharma, 'Material Background of the Genesis of the State and Complex Society in the Middle Gangetic Plains', in B. P. Sahu, ed, *Iron and Social Change in Early India, Debates in India History and Society*, New Delhi: Oxford University Press, 2006, p. 152. Compare with changes in landscape caused by rainfall patterns in Kalakamate, a village in north-east Botswana, see Annika Dahlberg and Piers M. Blaikie, 'Changes in Landscape or in Interpretation? Reflections based on the Environmental and Socio-economic History of a Village in NE Botswana', *Environment and History* 5: 2, 1999, pp. 127–74, particularly, pp. 147–49, 161–64.

[36] Sharma, 'Material Background of the Genesis of the State', pp. 150–68.

[37] Thapar, *Early India*, p. 44. Compare with river Terek in North Caucasus; see Barrett, 'The Land is Spoiled by Water', p. 39.

the meanders, the river accepts an average monsoon discharge of 1,500 m/sec with a high silt load and shifting areas of erosion and deposition.[38] This pattern can be applied to understand the swing and character of the Ganga prior to 1914 and perhaps even the earlier times.

History is thus evidence to the fact that the Ganga plain undergoes major flooding during the monsoon season, the pattern of which is influenced by the basic geomorphic build-up of the Ganga plain.[39] The climate and seasonality of rainfall in the Ganga plain directly controls the water budget of the river system and other water bodies. Consequently, the Gangetic rives and their water bodies receive the maximum water budget during the monsoons. One of the reasons for this could be that the climate of the region is controlled by the south-west monsoons.[40] We can therefore suggest, on the basis of the pattern noticed since early times, that the Gangetic rivers often got flooded and that their water bodies in turn flooded and inundated the adjoining low-lying areas.[41]

The floods of the Ganga, like all tropical rives, have produced quick alterations in the landscape. Extensive islands, rich in alluvium, were formed in the channel of the Ganga. Some of these islands, originally large sand banks, extended over four or five miles; they were formed at the angular turnings of the river, or in the straight parts or the middle of the stream. While the Ganga formed new islands in one part of its course, it swept away old ones in other parts. However, when the water subsided, the mould and mud collected on the island had a fertilising impact which facilitated thriving cultivation.[42]

Although the entire Ganga river system was prone to monsoonal floods, some of its tributaries like the Bhagirathi, Kosi and Gandak changed their courses more frequently and caused large havoc. Both Kosi and Gandak, in fact, have earned for themselves the title 'Sorrow of

[38] S. P. Das Gupta, ed, *The Upper Gangetic Flood Plain: A Regional Survey*, Calcutta: National Atlas Organization Monograph 1, 1975.

[39] Singh, 'Late Quaternary History of the Ganga Plain', p. 443; also V. S. Kale, ed, *Flood Studies in India*, Memoir 41, Geological Society of India, Bangalore, p. 229.

[40] Singh, 'Quaternary Palaeoenvironments of the Ganga Plain', p. 5.

[41] Compare with the valley of river Mississippi below Cairo over which the floods distribute themselves to a width exceeding 40 miles in many places; see Ballou, 'Floods, Their History and Relations', p. 1162.

[42] Rennell, 'An Account of the Ganges', pp. 100–01.

Bihar'. Notwithstanding the title, it may be observed that the impetuous Gandak has also affected some regions of eastern Uttar Pradesh. Later in this chapter, we will observe how the changing course of the Gandak influenced Gorakhpur (in the state of Uttar Pradesh) and became responsible for several land disputes in that region. Gandak, called Narayani river in Nepal, has changed its course over hundreds of years, from east to west. Every time it cuts its bank on the west, and especially during big floods, the territories of Nepal fall inside India. Till recent times, this has been a major cause of border disputes between India and Nepal. The Kosi has shifted its course westward by 112 km in 200 years, transforming 7,000–8,000 sq km of arable land to waste wetlands. Large amounts of sediment are transported downstream by the river and distributed across the fringing flood plains during the period of inundation.[43] Sweeping through the district of Purnia in north Bihar, Kosi increasingly became erratic in the nineteenth and early twentieth centuries.[44] Its violent flooding and meandering character rendered several areas unfit for cultivation, giving it the nickname 'River of Sorrow'.[45] Since the Kosi also marked the physical boundary between Bihar and Bengal, its shifting nature affected the demarcation of district Purnia. Consequently, in the pre-colonial period, the eastern portion of this district was considered part of Bengal and the western section, a portion of Bihar.[46]

Sixteenth-century Accounts of the Ganga

The turbulent nature of the Ganga is the subject of legend. According to the story, once Ganga came down to earth (from heaven), on its way to Bhagirathi, its rushing waters caused massive chaos and devastation,

[43] Amrita Chatterjee and Dipayan Dey, 'Water Woes in South East Asia: Geo-Ecology of Trans-Border River System and Dams between India and Nepal', accessed from http://balwois.mpl.ird.fr/balwois/administration/full_paper/ffp763.pdf on 3 May 2006.

[44] Christopher V. Hill, *River of Sorrow: Environment and Social Control in Riparian North India, 1770–1994*, Michigan: Association for Asian Studies, 1997, p. 5.

[45] L. S. S. O'Malley, *Bengal District Gazetteers: Purnea*, Calcutta, 1913, p. 6.

[46] Hill, *River of Sorrow*, pp. 17–18.

destroyed the fields and more significantly, broke the meditation of a sage called Jahnu. Annoyed by the violent flow of the Ganga, the sage drank up its entire water. It was only when the Gods prayed to Jahnu that he released the water of the Ganga from his ears. Since then the Ganga, also known as Jahnavi, has continued its tempestuous as well as beneficent course.

In the sixteenth century, or perhaps more appropriately in the period prior to the eighteenth century, the Ganga flowed through the Burh-Ganga (old Ganga) channel along the southern edge of the present *pargana* (village) Kharid (situated on the banks of river Saryu, now known as Ghaghara) in Bihar and joined river Ghaghara higher than it now does.[47] This is evident from Babur's (first Mughal emperor, 1526–30) estimate of the distance from Munir (in Bihar), situated on the bank of river Son and his camp (during his eastward campaigns towards Bengal and Bihar in 1527) on the Ganga. The distance measured then was 46,200 paces which is 12 *kurohs*, i.e. 24 miles.[48] The distance appears considerably greater now when compared to Babur's days. This was probably owing to the change of the river's course through its desertion of the Burh-Ganga (old Ganga) channel. The Burh-Ganga channel is now closely followed by the western boundary of pargana Ballia (in the present state of Uttar Pradesh).[49] Ganga and Ghaghara, during Babur's time, met where the Burh-Ganga entered the bed of Ghaghara.[50]

The *Babur Nama* also attests to the formation of lakes and ponds as a result of the changing course of the river and its channels. Describing his westward expedition in 1528 and 1529, Babur wrote that the army

[47] Zahiru'd-din Muhammad Babur, *Babur Nama*, trans. A. S. Beveridge, vols I and II, reprint, New Delhi: Oriental Books Reprint Corporation, 1970 (first published 1922) (henceforth *Babur Nama*); see especially vol. II, p. 674.

[48] *Ibid*., p. 667.

[49] When the Ganga flowed in the Burh-Ganga channel, the now pargana of Ballia was a part of the subdivision of Bihar. See *Babur Nama*, vol. II, p. 667n.

[50] During the march towards Bengal in 1527, Babur instructed his matchlock-men to pitch their camps, get their implements ready and engage in battle from the Bihar side of Ganga, which was a little below the meeting point of the Ganga and Ghaghara. Then again, during the preparation for war against Bengal, the raftsman, Mir Muhammad, represented that Saru (Ghaghara) was better to cross higher up. Accordingly the army was ordered to cross at the higher place. See *Babur Nama*, vol. II, pp. 668, 674.

lost its way and went to the great lake of Fathpur,[51] known as *Tal* (lake) Ratoi, which was then bigger and deeper than it is now. Subsequently, a second lake, larger than Tal Ratoi was formed further west. Through this tal, the Ghaghara once flowed and has constantly tried to break back into the tal.[52] These changes in the Tal Ratoi and in the course of the Ghaghara call for caution while attempting to locate the places mentioned by Babur.

Caution notwithstanding, it is not only Babur's account but even that of Bernier, a European traveller, who visited India in the seventeenth century, which confirms the shift in the course of rives, their impact on the environment and physical geography that remained transformed to a large extent. Bernier describes the formation of numerous islands in the gulf of Bengal at the mouth of the Ganga. The action of island formation was facilitated by the Ganga, carrying a great quantity of earth towards the sea. This pattern had rendered the Sunderbans a vast alluvial plain, where the process of land-making had remained continuous and ceaseless. It abounded in morasses and swamps and was intersected by rivers and estuaries running from north to south. These rivers and estuaries were connected with each other by innumerable smaller channels so that the whole tract became a tangled network of streams, rivers and watercourses, enclosing a large number of islands of various shapes and sizes.[53] The silt deposited in these islands by the rivers made them highly cultivable. Apart from this, the regular and excessive monsoonal rains obviated the necessity of cutting canals in Bengal for the purpose of irrigating the land.[54] While the Ganga appeared to be a bountiful stream, with exceptional utility for agriculture and navigation, its hydrograph was such that especially during heavy monsoons it caused floods and imbalances in the natural environment,

[51] Although read as Fathpur in *Babur Nama*, Beveridge suggests that Nathpur suits the context better than Fathpur. Abu'l Fazl's *Ain-i Akbari* (trans. H. Blochmann, second edition, New Delhi: Oriental Books Reprint Corporation, 1965) mentions Nathpur, a pargana in the Azamgarh district in the state of Uttar Pradesh. See *Babur Nama*, vol. II, p. 681n.

[52] *Babur Nama*, vol. II, p. 681.

[53] Francois Bernier, *Travels in the Mogul Empire, 1656–68*, trans. Archibald Constable, second edition, New Delhi: S. Chand & Co., 1968 (first edition, 1891), p. 452.

[54] *Ibid.*, p. 453.

even if not permanent damage or destruction to a region's ecology. It therefore became necessary to build embankments to restrain the river's inundation, for the alluvial silt which it spilled over its banks year after year was an inexhaustible source of fertility to the fields. If one crop was drowned by the floods, the cultivator could calculate that his second crop would abundantly requite him.[55]

The voluminous discharge of the Ganga and its tributaries and their rising and raging currents during heavy rains were a matter of grave concern to the Mughal emperors: particularly during their military expeditions if the army had to cross the river, often carrying artillery and guns, or pitch their tents on the banks of the river.[56] Babur, for instance, considered it advisable to plan battles subsequent to the monsoons, especially if the expedition entailed crossing the Ganga. This would imply that there was a constant threat of the Ganga and its tributaries swelling during the monsoons, arousing fears of inundated banks and changing river channels. In this context, we may cite an event from Babur's period. In 1528, the various rebellious *amirs* (nobles) of the east, i.e. Nasir Khan Nuhani and Ma'ruf Farmuli, had crossed the Ganga, occupied Kanauj and were merely 3 miles away from Babur's camp on the river. Yet, apprehending the overflow of Ganga during the rains, Babur showed no counter reaction. Even when Rana Sanga (the ruler of Mewar in Rajasthan) captured Qandhar and maintained a 'hostile and mischievous attitude', Babur remained silent.[57] His calmness may be read as expectation that the rain would wane enabling him to prepare counter-attacks. As the rains approached their end, Babur found it expedient to move against the rebels and Rana Sanga.[58] On another occasion in June 1529, Babur narrates about the overflowing Yamuna and its impact. Describing his westward expedition, he says that as his army dismounted at pargana Adampur, situated on the west bank of river Jun (Jumna or Yamuna), they found Yamuna in flood, and by dawn the whole of the island on which the army had pitched its camp was under water. They thus had to move their tents to another island 'an arrow's flight downstream'.[59]

[55] *Ibid.*

[56] Zain Khan, *Tabaqat-i Baburi*, trans. S. Hasan Askari, New Delhi: Idarah-i Adabiyat-i Delli, 1982, p. 150–51.

[57] *Babur Nama*, vol. II, p. 580.

[58] *Ibid.*

[59] *Ibid.*, p. 685.

Even Abu'l Fazl, a chronicler of the sixteenth century during the reign of Akbar (the third Mughal emperor, 1556–1603) gives us an exhaustive account of the impetuous Ganga and its tributaries. Describing the capture of the city of Patna by Akbar in August 1574, Abu'l Fazl observes that the emperor chose the river route despite the 'constant rain and tempest... and season full of turbulence...'.[60] Patna and its suburbs were reportedly inundated and the military reinforcements including horses and arms and ammunitions were washed away. Munim Khan, the governor of Bihar, had allegedly petitioned Akbar to order the recovery of damages caused by the floods.[61] As Akbar approached Hajipur (the fort opposite Patna), he realised that the Ganga which was about 2 *kos* broad between the two cities 'flowed with great force and turbulence'.[62] However, undeterred by the furious flow of the river, soldiers under the command of Khan Alam mounted river-traversing camels, crossed the river with suitable artillery and captured the forest of Hajipur (which was seen as a means of subduing the city of Patna).[63] Similarly, when the imperial army of Akbar marched towards Bengal in September 1574, it was rainy season. As a result, the movement and encampment of the army were made difficult since most regions of Bengal were under water. The flooding of Bengal, Abu'l Fazl notices, was owing to the several torrential rivers which join Ganga in Bengal and cause 'typhoon-like violence'.[64] In 1584–85, floods swept the delta of river Meghna, devastating the economic prosperity of Akbar's reign.[65] Apart from Bengal, Abu'l Fazl also notices inundations in *sarkar* Sambhal (Rohilkhand, a region in the state of Uttar Pradesh) caused by the river Saryu (Ghaghara).[66] That the

[60] Abu'l Fazl, *The Akbar Nama*, vol. III, second edition, trans. H. Beveridge, New Delhi: Rare Books, 1973, p. 135 (henceforth *Akbar Nama*).

[61] Augustus Frederick, *Kaiser Akbar*, trans. A. S. Beveridge, Calcutta: Thacker, Spink & Co., 1890, pp. 224–25 cited in Agrawal, *Great Natural Calamities*, p. 59.

[62] *Akbar Nama*, vol. III, p. 135.

[63] *Ibid.*, p. 136.

[64] *Ibid.*, p. 645.

[65] S. S. Kulshreshtha, *The Development of Trade and Industry under the Mughals, 1526–1707*, Allahabad: Kitab Mahal, 1964, p. 33.

[66] Fazl, *Ain-i Akbari*, vol. I, pp. 303, 403, cf. Irfan Habib, *The Agrarian System of Mughal India, 1556–1707*, second edition, New Delhi: Oxford

pattern of flooding in Bengal had remained almost the same even till early nineteenth century is evident from James Grant's comments on the low-lying lands of Bengal. He observes that the low marshy land of Hijli, anciently called Bhati,[67] were to a great extent subject to the overflowing of tides from the Gangetic delta in Bengal.[68]

The threatening hydrograph of Ganga and its tributaries could not however restrain the Mughal emperors. On different occasions, in 1526, 1528 and 1529, Babur ordered the construction of bridges over the Ganga to facilitate the army in crossing it and carrying their guns.[69] We also have references that during the Bengal campaign of 1574, Akbar built bridges across the Ganga for the imperial army to proceed further. To escape the ravages of the Ganga, when there was rain and tempest, the soldiers and sometimes even the rebels crossed the *juibar* (canal), cut at the head of the Ganga. Abu'l Fazl observes that this channel was cut from the river, sometime during the ancient period. This cutting was probably meant to be an inlet from the old Ganga or Bhagirathi, leading towards Tanda.[70]

University Press, 1991, p. 31. Saryu, which now meets Ghaghara in the south-eastern corner of Kheri district, used to flow at the time of *Ain-i Akbari*, past Bahraich and joined Ghaghara at only one *kuroh* above the town of Ayodhya. It thus had a longer course in the plains than now. The old channel can still be traced on the maps. See Habib, *The Agrarian System of Mughal India*, p. 31n.

67 Beveridge observes that Abu'l Fazl regarded Bhati as lying on the east as well as the south of Bengal. He probably regarded the whole of Sylhet, the southern part of which is very low, as belonging to Bhati. See *Akbar Nama*, vol. III, p. 646n.

68 James Grant, 'Analysis of the Finances of Bengal from the Mogul Conquest to the Present Time', Calcutta, 30 June 1787, in W. K. Firminger, *Affairs of the East India Company* (*Being the Fifth Report from the Select Committee of the House of Commons*), 28 July 1812, vol. II, reprint, New Delhi: Neeraj Publishing House, 1984.

69 *Babur Nama*, pp. 495, 599, 633.

70 *Akbar Nama*, vol. III, p. 443n. Tanda, locally called Tarrah, was the then capital of Bengal. The Ganga bed between Tanda and Gaur was apparently the old bed of the Ganga and was known as Bhagirathi. It is believed that the Bhagirathi had once washed away Tanda in the mid-sixteenth century. See *Akbar Nama*, vol. III, pp. 428n, 442n; also see Charles Stewart, *History of Bengal: From the First Mohammadan Invasion until the Virtual Conquest of that Country by the English, AD 1757*, London: Black, Parry & Co., 1813, p. 149n.

In 1658, the inaugural year of the reign of Aurangzeb (the sixth Mughal emperor, 1658–1707), the flow of the rivers Ganga and Yamuna was so intense at Allahabad (a town in the state of Uttar Pradesh) that the entire city except the fortress of Allahabad was submerged under water, drowning several people. Referring to such disasters, Manucci, an Italian traveller, observes that disasters of this kind were a signal for impending wars, calamities, destruction and miseries for the people.[71] Again, in 1691, the furious flow of the river Bhima (where Aurangzeb had pitched his military camp) inundated its banks, wiping out thousands of men and their homes, and destroying and washing away imperial establishments like horses, bullocks and cattle. The disaster was so severe that to alleviate the miseries of the people, Aurangzeb propitiated God, believing that divine intervention alone could control the catastrophe. He wrote out prayers and ordered them to be thrown in the waters of river Bhima so that its ferocity and intensity could subside.[72] In invoking divinity, Aurangzeb emulated the thinking of his predecessor, Babur. Babur, too, believed that the violence of rains and floods could be mitigated by prayers and magic. He had apparently taught Mulla Ali Khan the use of a talisman which could stop rain.[73]

Strategies to Cope with the Ganga in Spate

The stormy hydrograph of the river Ganga has had a tremendous impact on the economy and livelihood of the regions through which it flowed. The Ganga, as it passed through district Purnia in north Bihar, was rendered extremely calamitous by the Kosi and Ghaghara rivers. It has been observed that no part of the inland navigation of India was as dreaded or dangerous as the Ganga at its confluence with Kosi.[74] The Ganga has frequently changed its course within the district and formed

[71] Niccolao Manucci, *Storia do Mogor*, vol. II, trans. William Irvine, New Delhi: Atlantic Publishers and Distributors, 1990, p. 428.

[72] Mustaid Khan Saqi, *Maa'sir-i 'Alamgiri, 1558–1707*, trans. J. N. Sarkar, Calcutta: Royal Asiatic Society of Bengal, 1947, pp. 236–37; also Khafi Khan, *Muntakhab-ul-Lubab*, vol. II, p. 450 cited in Agrawal, *Great Natural Calamities*, pp. 64–65.

[73] *Babur Nama*, vol. I, p. 426.

[74] O'Malley, *Bengal District Gazetteers*, p. 7.

huge areas of *diara*[75] along the southern border. The constant process of erosion and deposition, however, has made the diara abundantly fertile and has therefore encouraged a high density of population.[76]

In Bengal, the Ganga and its distributaries have swept away vast tracts, changing their identities completely. The raging spate of the Ganga has, in fact, rendered the physical environment of Bengal so transient that the peasants have almost resigned themselves to its vagaries. Their desire for security and the wish to cross the river peacefully is evident from this popular folk song: *Hari din to gelo, shondhe holo, par karo amare* (The day is done and the evening come, pray my God, take me over the ferry).[77]

The sentiments expressed in this song suggest that the river Ganga has changed its course several times and disturbed the balance of the entire river system. These periodic changes were perhaps caused by gradual levelling of the land and silting of the old river beds.[78] Since the early seventeenth century, the Ganga has shown a constant shift towards the east. This eastern swing was further enhanced by the environmental upheavals of the later eighteenth century. The earthquake of 1762 and the floods of 1769–70 and 1786–88 created 'new' rivers and gave a new identity to old rivers. These 'new' rivers shaped by natural calamities of the eighteenth century include Tista, Jamuna, Jelanghi, Mathabhanga, Kirtinasa and Naya Bhangini.[79] Subsequent to the floods of 1787, the Tista, earlier a tributary of Ganga, joined the Brahmaputra, which by the eighteenth century had moved westwards to meet the Ganga. The eastward incline of the Ganga and westward shift of the Brahmaputra changed the ecological context, bringing to Bengal bounties by facilitating agriculture and developing an agrarian society.[80] The constantly changing course of the river has created

[75] *Diara*, i.e. fluctuating river tract, comprises the land that a river leaves behind or that it inundates annually and the land bordering the new beds that it may form during the course of its movement. See Hill, *River of Sorrow*, p. 5n.

[76] Hill, *River of Sorrow*, p. 19.

[77] Bose, *Peasant Labour and Colonial Capital*, pp. 9–11.

[78] B. N. Ganguli, *Trends of Agriculture and Population in the Ganges Valley: A Study in Agricultural Economics*, London: Methuen & Co. Ltd., 1938, p. 205.

[79] R. K. Mukherjee, *The Changing Face of Bengal: A Study in Riverine Economy*, Calcutta: University of Calcutta, 1938, p. 9.

[80] Bose, *Peasant Labour and Colonial Capital*, p. 12.

two remarkably contrasting deltas—an active delta in the east and the moribund delta in West Bengal. The eastern delta enjoys the fertilising impact of the floods and the overflowing rivers and is often referred to as the 'land of the new alluvium'.[81] The fact that the two areas are vastly and qualitatively different is evident from their cropping patterns and demographic status.

It is evident from our discussion that the turbulent Ganga had aroused fears of sorts for some regions and some individuals while others may have been the beneficiaries of it. In this context, we may mention instances when the capricious Ganga had kept the East India Company on tenterhooks on several occasions. The ordnance factories and bungalows of the company on the banks of the river in Patna and Bankipur were at high risk of being swept away or destroyed during the monsoons. To secure the banks of the river in front of the Patna Magazine, against the force of the river's current in the rainy season the company had built embankments of brushwood, bamboo, sal timber and small timber.[82] But in 1783, the spate of the river was so high (much above the height of the bank which was 29 ft) that it encroached 10 ft on the Patna Magazine, swept away the embankment of brushwood, bamboo and small timber and destroyed 10 ft of the solid bank, although it spared the large sal timber. It appears that sal timbers alone were capable of withstanding the rapid, rushing current of the river.[83] However, they needed to be preserved. With the annual

[81] *Ibid*. Compare with the political division known as *Mughalbandi* or alluvial plains in Orissa, see Rohan D'Souza, *Drowned and Damned: Colonial Capitalism and Flood Control in Eastern India*, New Delhi: Oxford University Press, 2006, pp. 52–53.

[82] John Macintyre, Secretary to the Board of Ordnance to W. Bruere, Secretary, Governor General-in-Council, Ordnance Office, 15 April 1784, Original Consultation, 19 April 1784, No. 28, Home Department, Public Branch, 1784 (National Archives of India, New Delhi, henceforth NAI). Compare with the embankments in Orissa, which were inspired by the 'protective' embankments of Bengal to separate the cultivated fields from the river spill. The records of Orissa suggest that such embankments had been built much before the establishment of the Company rule in the region; see D'Souza, *Drowned and Damned*, particularly the chapter titled 'Embankments and its Discontents', pp. 97–125.

[83] Robert Lands, Commisary of Ordnance at Patna to John Macintyre, Secretary to the Board of Ordnance, Patna Magazine, 15 March 1784, Original

gush of the river, the sal timber decayed at the end of seven or eight years, weakening the roots and making them susceptible to collapse and being washed away during the next heavy monsoons. The fascines made of brushwood, turin and bamboo could not however last for more than two years. They rot with the percolation of water to their roots during the second season of rains and had to be replaced every third season of the rains to prevent the fascines from being swept away. Despite these risks, not only the company but even the local inhabitants of Patna used embankments of brushwood, bamboo and turin to preserve their houses.[84]

The floods of 1787 had wiped away the embankment of brushwood, bamboo and small timber, which protected the company's bungalow at Patna and Bankipur. It was feared that unless the bank of the river was secured by piles and fascines, a part of the bungalow would be carried away in the next rains.[85] The violence of the floods had reduced the *ghat* (bank) at the upper end of the river to 'pieces' and removed a round bastion built opposite the company's house in Patna. To save the building from the encroaching river, it was proposed to either face the bank with sal timber or fascines of brushwood or build a fort-like substantial wall on strata of *kanker* (stone) rock to provide the wall the solidity and sustainability to resist the furiously running waters.[86]

The threat to the ordnance factory at Patna was even more severe with one corner of the factory standing at the distance of a yard from the edge of the river bank. Since the focus of the company was

---

Consultation, 19 April 1784, No. 29, Home Department, Public Branch, 1784 (NAI).

[84] Letter from Captain J. Garstin to Lt. Col. Thomas Call, Chief Engineer, Patna, 1 January 1788, Original Consultation, 14 January 1788, No. 11, Home Department, Public Branch, 1788 (NAI). Compare Duchy of Ferrara in North Italy, where embanking of rivers by residents was obligatory for territorial water control otherwise they were fined for non-compliance. See Guidoboni, 'Human Factors, Extreme Events and Floods', p. 292.

[85] Extract from a letter, 17 November 1787 from the Resident at Patna to the Board of Revenue, Original Consultation, 6 December 1787, No. 9, Home Department, Public Branch, 1787 (NAI).

[86] Letter from Captain J. Garstin to Lt. Col. Thomas Call, Chief Engineer, Patna, 1 January 1788, Original Consultation, 14 January 1788, No. 11, Home Department, Public Branch, 1788 (NAI).

on preserving and protecting the cantonment, there was a growing anxiety to check further encroachment and manage the course and rapidity of the river. The bank opposite the cantonment was therefore kept well sloped so as to oppose an inclined plane to the streams.[87]

None of these strategies could, however, save the large round bastion built to secure the town of Bankipur from the turbulent Ganga. The purpose of the bastion was to throw the force of the river towards the mouth of the Gandak. But with the fallen bastion, the strength of the current which was earlier diverted now exerted its whole force on Bankipur. Consequently, the company and the local inhabitants who possessed any property on the banks of the river were exposed to the stormy river and its encroachment and also spent a considerable sum on securing the bank. These hazards notwithstanding, the washed away round bastion was not a disaster for all. The land on which it stood was transformed into a vast island, fertilised by the silt deposited by the overflowing river. Although the silt and sand did not remain constant and kept shifting their position, the island was rendered highly cultivable.[88] This process may corroborate the general pattern of urbanisation seen in the Gangetic plain over a period of nearly 2,000 years. Between the settlement and the regular course of the river, there was a strip of land of roughly 150–300 m, but this space widened in some places after the monsoon floods. This strip was fertile and rich in organic matter.[89] However, if one draws a circle around the settlement, one-third of it would go into the river. Therefore, not more than two-thirds was available for cultivation.[90]

[87] Copy of a letter from E. E. Pote, Resident at Patna to Charles Stuart, President and Member of the Board of Trade, Singia Factory, 25 January 1788, Original Consultation, 13 February 1788, No. 4, Home Department, Public Branch, 1788 (NAI); also see, Letter from Captain J. Garstin to Lt. Col. Thomas Call, Chief Engineer, Patna, 28 April 1788, Original Consultation, 19 May 1788, No. 5, Home Department, Public Branch, 1788 (NAI).

[88] Letter from Captain J. Garstin to Lt. Col. Thomas Call, Chief Engineer, Patna, 28 April 1788, Original Consultation, 19 May 1788, No. 5, Home Department, Public Branch, 1788 (NAI).

[89] Makhan Lal, 'Iron Tools, Forest Clearance and Urbanization in the Gangetic Plains', in B. P. Sahu, ed, *Iron and Social Change in Early India*, p. 145.

[90] *Ibid.*

River Gandak in Gorakhpur

The tributaries of the Ganga—Gandak, Ghaghara, Rapti (and their tributaries) flowed through Gorakhpur (Uttar Pradesh). These rivers largely had a fertilising impact and they enhanced the productivity of the soil by the alluvium which they carried. However, like all rivers, there was actually nothing permanent in these rivers for very little endured in their deltas. They carried away soil with their torrents, or altered their course leaving behind beds or islands or spaces, either immediately cultivable or subsequently becoming so. They often overflowed their banks, subjecting them to frequent inundation through they also restored and recuperated them with their rich alluvial deposits. This pattern notwithstanding, in some instances, inundated lands were rendered completely unproductive and barren.[91]

The frequently changing course of river Gandak had created *gangshikisht* and *gangberamud* lands, meaning lands carried away or thrown away by the changing course of the rivers. In Gorakhpur, there were innumerable disputes over these lands. The conflict over this issue began when the *zamindars* (landlords) of the adjoining parganas or districts claimed ownership rights over gangshikisht and gangberamud lands, which meant that the zamindar whose lands were washed away by the flow of the river claimed possession over it by virtue of his original right, but at the same time the zamindar to whose territories the new lands were ceded also asserted his rights over it.[92] The impetuous nature of river Gandak and its tributaries made the zamindars even more rebellious and belligerent. To settle such discord, the East India Company often resorted to customary laws. But the legal code—both Hindu and Muslim—had laid down no exclusive private rights in the lands formed by alluvium. Nevertheless, long usage and traditions of the villages along the coast of the river confirmed a title of possession to the water's edge. This implied that any new land formed by the changing course of the river belonged to the landholders on the bank where this occurred, though the river by the alteration of its course

[91] Bhargava, *State, Society and Ecology*, pp. 10–11.

[92] J. Ahmuty to Captain John Baillie, Resident at Lucknow, 9 September 1808, Boards Collections, 1808–9, vol. 251, F/4/251, 5633 (The British Library, London, henceforth BL).

may have inflicted losses on the original proprietor by washing away his lands. Even with regard to the islands, beds or spaces formed by the rivers, there was no unanimity over the ownership. Some believed that the landholders on either side may claim a right over these formations, others argued that they belonged to the proprietor on the nearer shore and yet others suggested that if the current of the river did not flow between the island and the shore, leaving the space dry, then the island belonged to the landholder on the dry shore.[93]

The lands of Ramnarain Rai and Bahadur Rai, the zamindars of several villages in pargana Padrauna, had been broken and carried away by the strides of river Gandak. The river had thrown away 14 cultivated and waste villages together with the forests containing sesoo timber and long pepper to the other side of the river, attaching them to the estates of Raja Bir Kishore in sarkar Champaran, district Saran (now in Bihar).[94] Raja Bir Kishore, in accordance with long usage and tradition, claimed possession over these lands. The zamindars of Padrauna, on the other hand, demanded compensation for their lands. The district collector rejected their plea but gave them the option of filing a suit against Raja Bir Kishore. Meanwhile, within the year, the erratic Gandak reinstated the formerly wiped out lands to Padrauna, the zamindari of Ramnarain Rai and Bahadur Rai. This initiated the endless tussle between the zamindars of Padrauna and Champaran, both averring their rights to gangshikisht and gangberamud lands.[95]

On yet another occasion, the deviations of river Dewa, a tributary of Gandak, aroused persistent conflict between Raja Ram Prakash Singh, zamindar of pargana Aurangabad Nagar and the *amils* (revenue collectors) of Tanda (on behalf of the Nawab of Awadh). The dispute occurred over an island, a portion of the land of the zamindar of pargana

[93] Collector Banaras to T. Fortescue, 18 April 1809, Proceedings Board of Commissioners—Ceded and Conquered Provinces (henceforth Procs. Bd. Com. CCP), vol. 18, 1809 (Uttar Pradesh State Archives, Lucknow, henceforth UPSA).

[94] Petition of Ramnarain Rai and Bahadur Rai, zamindars vs. Collector Gorakhpur, 9 May 1810, Enclosure in Charles Buller to F. Balfour, 3 June 1810, Procs. Bd. Com. CCP, vol. 35, October 1810 (UPSA).

[95] F. Balfour to T. Graham, 17 April 1807, Proceedings Board of Revenue—Fort William, vol. 32, April 1807 (UPSA).

Aurangabad Nagar, which was washed away by river Dewa to the side of Tanda. In accordance with the local custom, the zamindars of Tanda claimed possession to it. Subsequently, when the river re-annexed the island to pargana Aurangabad Nagar, the dispute worsened with Ram Prakash Singh demanding the restoration of his rights.[96]

These disputes—whether between Padrauna and Champaran or Aurangabad Nagar and Tanda—illustrate that several zamindars across Gorakhpur and particularly in regions affected by the impetuosity of the Gandak and its tributaries, remained engaged in a continuously recurring discord.

In our discussion of the changing course of the river Ganga and its tributaries, we have identified the turbulent hydrograph of the rivers which have caused natural disasters in the form of floods, tides and inundated banks. Erosion (the silting up of one bank while the other is being eroded) and avulsion (the cutting of a new channel) have been its major characteristics and method of altering its course. Yet its long historical past has demonstrated its enduring nurturing qualities, its self-regulating mechanism of recuperating and reviving the areas it had inflicted catastrophe upon. This, however, does not discount the permanent damages and destruction that it may have caused to a region's ecology or that its bounties have increasingly proved inadequate for the sustenance of human beings.

[96] J. Ahmuty to Captain John Baillie, Resident at Lucknow, 9 September 1808, Boards Collections, 1808–9, vol. 251, F/4/251, 5633 (BL).

Section IV
THE POLITICS OF ANIMALS

The Elephant in Medieval Indian Warfare*

Simon Digby

The domestication and use in warfare of the elephant has since Carthaginian times been almost confined to Greater India, with rare and tactically insignificant importations into the Middle East and other areas. In India itself, the elephant was employed in warfare at least since the fourth century BCE and was probably domesticated some centuries before that. Improved musketry rendered its employment in battle almost obsolete in the seventeenth century, when elephants were principally used to mount generals or commanders, in order that they might be seen by their troops and obtain an extended view of the battle; even this had its disadvantages and Nadir Shah is alleged to have remarked: 'What strange practice is this that the rulers of Hind have adopted? In the day of battle they ride on an elephant and make themselves a target for every one.'[1] When wounded, elephants were liable to get out of control and escape at the top of their speed. Irvine cites the case of 'Azim al-shan, contender for the Mughal throne in 1712: his elephant when wounded jumped off a high bank into the river Ravi, drowning the prince along with himself.[2]

The elephant had incidental uses in medieval warfare outside the battle line. Heavily armoured, it could be used as a living battering ram to push down the gates of a fortress. Many of the strongest fortresses in India have elephant spikes upon their doors to hinder this form of assault.[3] In one instance in the period of the Delhi Sultanate we read

* Simon Digby, 'The Elephant in Medieval Indian Warfare', in Simon Digby, *War-Horse and Elephant in the Delhi Sultanate: A Study of Military Supplies*, Oxford, Orient Monographs, 1971, pp. 50–82.

[1] Dalpat Ray, *Malahat-i-Maqal*, quoted by W. Irvine, *The Army of the Indian Moghuls*, London: Luzac & Co., 1903, p. 177.

[2] *Ibid.*

[3] Sidney Toy, *The Strongholds of India*, London: William Heinemann, 1957, pp. 6–7: also Toy, *The Fortified Cities of India*, London: Heinemann Educational, 1965, p. 24.

of elephants being brought against the gate and the fortress falling.[4] The elephant could also serve as a pack animal carrying a very large indivisible load, for instance a royal tent, and is recommended by Sir George Watt, writing at the close of the nineteenth century, as a means of transport through dense and impenetrable country.[5] In 'Afif's account of one of the campaigns of Feroz Shah Tughluq we read of the solution to the problem of how to ford a river flowing so swiftly that men and horses were liable to be borne away. The elephants were deployed in two chains, attached to one another by ropes, and were made to stand upstream and downstream of the ford. The chain of elephants upstream broke the force of the rushing waters, and the chain downstream served as a net, against which horses and riders who had been carried off by the water were caught instead of being swept away altogether.[6]

Nevertheless elephants were principally esteemed and treasured in medieval India for their performance upon the field of battle. About this Professor Basham speaks with less than his habitual charity: 'The great reliance placed on elephants was, from the practical point of view, unfortunate... Even the best trained elephants were demoralized comparatively easily... The pathetic Indian faith in the elephants' fighting qualities was inherited by the Muslim conquerors, who, after a few generations in India, became almost as reliant on elephants as the Hindus and suffered at the hands of armies without elephants in just the same way.'[7] In the Delhi Sultanate, the only army with elephants which suffered at the hands of a foreign and elephant-less army was that of 1398 CE. The most apparent cause of its defeat was its pathetically diminished numbers both of horses and of elephants, compared with the armies which the Sultans of Delhi had put into the field during earlier invasions.

[4] Muhammad Bihamad Khani, *Ta'rikh-i Muhammadi* (henceforth *TMH*), MS. British Museum Oriental 137, folio 426A: *pil u lashkar bar dar-i qal'a-i Kotra burd.*

[5] George Watt, *Dictionary of the Economic Products of India*, vol. III, London: W. H. Allen & Co., p. 221.

[6] Shams ud-din bin Siraj ud-din 'Afif, *Ta'rikh-i Ferozshahi* (henceforth *ATFS*), M. Vilayat Husain, ed., Calcutta: Asiatic Society of Bengal, 1891, p. 111.

[7] Arthur Llewellyn Basham, *The Wonder that was India*, London: Sidgwick & Jackson, 1954, pp. 129–30.

The elephant is a picturesque animal, and medieval authors are all convinced that it was a great asset in battle. Examples of its performance on the battlefield during the Sultanate period do not decisively support this view, and we must allow for the aesthetic enthusiasm which the elephant, like the horse, evoked: '... a meet seat for a king and a servant of the court, its body was heavy and its paces were gentle; it could break the enemy lines yet fight in ordered ranks. "And when they were together in a row, there is an earthquake of *Fad! Fad!* and *Saf! Saf!*"'[8]

Al-'Umari describes the 3,000 elephants of the Delhi Sultan. They had litters or howdahs upon them and were covered in wartime with gilt or perhaps inlaid (*mudhahhab*), steel animal armour (*barkustuvan*), in peacetime with *dibaj*, *vashi* and other types of silk. The superstructure of their howdahs consisted of a broad back-plate or platform to which was nailed a tower of wood. The Indian drivers rode in front of these. The elephant could carry six to 10 men according to its size.[9] Later al-'Umari describes the employment of the elephant in battle. The Sultan in the *qalb* or centre of his army would be surrounded by *a'imma* and *'ulama*: before him and behind there would be archers. To the left and the right, without any break, were the wings. The elephants stood in front of the Sultan, clad in steel armour and upon their backs the covered towers with warriors inside. These towers had apertures through which arrows were fired and flasks of Greek fire (*qawarir al-naft*) were thrown. In front of the elephants stood foot-soldiers in a small troop to clear a path for the elephants and to ward off the enemy cavalry before they reached the elephants; and these foot-soldiers were themselves covered by the archers or fire-throwers in the towers on the elephants. Al-'Umari once again in the same passage refers to the arrows and the *naft* launched from the elephants.[10]

Other references support al-'Umari's statement that the elephants were placed in the front rank and generally in the centre, although at the battle of Kili in 1299 CE there were evidently considerable numbers

[8] Amir Khusrau Dehlavi, *Khaza'in al-Futuh* (henceforth *KF*), Muhammad Wahid Mirza, ed., Calcutta: Asiatic Society, 1953, p. 107; English Translation by Muhammad Habib, Madras: Diocese Press, 1931, p. 76.

[9] Ibn Fazl Allah al-'Umari, *Masalik al-absar fi mamalik al-amsar* (henceforth *Masalik*), portion concerning India, Khurshid Ahmed Fariq, ed., New Delhi: Kitab Bhawan, 1961, pp. 24–25.

[10] *Masalik*, pp. 52–53.

of elephants—'two hundreds' according to 'Isami—at the wings.[11] However the Indian elephant is capable of what has been called 'a fast shuffle of about fifteen miles an hour'[12] and they seem on occasions to have been brought to charge, when it would have been almost impossible to keep them in their protective phalanx of foot-soldiers in the manner which al-'Umari describes. The poet Amir Khusrau describes the elephants closing into battle:

> The rank of elephants was like a line of baneful clouds,
> Each cloud with lightning to attack, swift like the wind,
> In its swift motion each elephant like a splendid mountain,
> The armour upon it like the cloud upon the mountain;
> On the back of the elephant Turks with arrows by their thumb-rings,
> Like a mountain which has seated itself on a mountain;
> Behind the elephants the cavalry with their line drawn out
> In frenzy drawing spume from the back of the cosmic fish;
> Not a single rank, nay rather a hundred weighty ramparts
> From which the empty space of the world became crowded.
> In the midst of the Centre, the *murtadd* with his umbrella over his head,
> Beneath the (royal) umbrella a little moist toadstool.[13]

Amir Khusrau was writing from the point of view of the other side, who won the battle without the aid of elephants, slaying one elephant and capturing 12. He describes the elephants in defeat:

> From arrows the elephant was grafted with arrow-notches
> Like a porcupine with its back full of quills:
> From the elephant its driver was hanging,
> His body hanging and his life fled:
> Other drivers were endeavouring and hurrying
> To make the elephant enter an ant-hole.[14]

As in this battle, in the disaster of 1398 CE very few elephants appear to have been killed on the battlefield.[15] The largest proportion noted

[11] Isami, *Futuh al-Salatin* (henceforth *FS*), A. S. Usha, ed., Madras: University of Madras, 1948, p. 260.

[12] Watt, *Dictionary of the Economic Products*, vol. III, p. 219.

[13] Amir Khusrau Dehlavi, *Tughluq-Nama* (henceforth *TUN*), Syed Hashmi Faridabadi, ed., Aurangabad: Matba Urdu, 1933, p. 92.

[14] *TUN*, p. 98.

[15] *TUN*, p. 81.

of elephants killed in battle is in 'Isami's description of the battle of Kili in 1299 CE where he has mentioned 'two hundreds' of elephants drawn up in the wings of the Sultan's army[16] and states that 30 were killed when Zafar Khan with the Right (*maymana*) was encircled by the Mongols.[17]

Barani's descriptions of the same incident illustrate both the advantages and the disadvantages of the employment of war-elephants. The Mongol Left gave way and Zafar Khan and the Sultan's Right pursued them, unsupported, for 18 *kos* into what may have been an ambush. Zafar Khan and his force perished after putting up a desperate resistance with their arrows. His elephants were wounded and the drivers killed.[18] The variant recension adds the detail: 'and six or seven elephants which were in front of Zafar Khan suffered severe wounds: they threw their drivers and came back safely.'[19] In this case the elephants, when they were out of control, had broken through the ranks of the encircling enemy and presumably carried some at least of those who were in their howdahs to safety.

The Numbers of Elephants in the *Pil-khana*

Perhaps our strictures with regard to medieval statistics are less true with regard to elephants than with regard to either horses or men. The numbers of elephants being so much smaller, the individual elephant is of greater importance. Amir Khusrau, writing an official account of the great Deccan campaigns of Malik Kafur, could devote space to the story of the pursuit and capture of three elephants which had been withheld when the Ray of Arangal had surrendered his total store.[20] 'Afif describes with laudatory phrases the hunt of Feroz Shah Tughluq in which seven wild male elephants and one particularly

[16] *FS*, p. 260.

[17] *FS*, p. 267.

[18] Ziya al-din Barani, *Tarikh-i Firuzshahi* (henceforth *BTFS*), Syed Ahmed Khan, ed., Calcutta: Asiatic Society of Bengal, 1862, pp. 260–61.

[19] *BTFS-A*, A manuscript of a variant recension of *Tarikh-i Firuzshahi*, representing an earlier draft of the author, Writer's Collection, Bodleian Library, Oxford, Accession No. 57, folio 97 A.

[20] *KF*, p. 88.

savage female were captured.[21] Occasionally a mistranscription or slip of the memory can occur, as when Barani says that 612 elephants were passed in review before the Sultan as spoils of Malik Kafur's southern expeditions,[22] whereas Amir Khusrau, who is probably his source, says 512.[23] But it is on the whole relatively difficult to doubt such small and circumstantial figures.

The Ghaznavid kingdom had employed war-elephants on a considerable scale. Mahmud of Ghazna inspected 1,300 elephants at the muster of 1023–24 CE and his son Mas'ud 1,670 elephants in 1031 CE. As Bosworth remarks, the last number agrees with that mentioned in a contemporary verse of Farrukhi: 'One may ask, "What are those 1,700 odd mountains?" I reply, "They are the 1,700 odd elephants of the Shah"'[24] Compared to the numbers of war-elephants mentioned in the Delhi Sultanate, these figures for the Ghaznavid kingdom seem surprisingly high, but it is difficult to doubt their authenticity. Even the court poet Farrukhi, presenting a *qasida* on a specific occasion, would be unlikely to congratulate the Sultan on having 1,700 elephants if the real figure was nearer, say, 700: though he might round off 1,670, the figure mentioned by the historian Bayhaqqi, to 1,700.

It is possible that the north Indian breeding grounds of the elephant may have been diminished by over-exploitation or by settlement and cultivation between the early eleventh century and the thirteenth. This would account for a relatively higher number of elephants in the Ghaznavid *pil-khana* (elephant stables). Some support is given to this idea by the numbers recorded by the Ghaznavid historians as captured from individual north Indian rulers: '350 from Qanawj and 150 from Mahaban in 1018–9 and 580 from the Raja Ganda in 1019–20.'[25] However the detailed information from the early sixteenth century which we possess regarding the capture of elephants in northern India seems rather against this hypothesis.[26]

[21] *ATFS*, p. 168.

[22] *BTFS*, p. 333.

[23] *KF*, p. 161.

[24] Clifford Edmund Bosworth, *The Ghaznavids*, Edinburgh: Edinburgh University Press, 1963, pp. 116–17.

[25] *Ibid.*, p. 116.

[26] 'Abd Allah, *Ta'rikh-i Da'udi*, Shaikh Abdur Rashid, ed., Aligarh: Aligarh Muslim University, 1954, pp. 30–31.

The numbers of the war-elephants of the Delhi Sultanate may also be compared with later figures of elephants employed in warfare in India. Two high figures for the fifteenth century should be treated with caution. Sultan Mahmud Sharqi is said to have marched against Bahlul Lodi in 1452 CE with 1,400 war-elephants (*fil-i jangi*) and 170,000 horse and foot. However the figures are those of an early seventeenth-century source of no great liability.[27] Farishta quotes from a lost contemporary chronicle of the Bahmani rulers of the Deccan, which writes of the exceptional number of 3,000 elephants which were collected in the pil-khana of Muhammad III Bahmani (r. 1463–82 CE).[28] However this description specifically states that this was the number of elephants of all kinds, big and small, male and female, in the pil-khana. Only a portion of these, perhaps a quarter, would be war-elephants, a large, but not improbable number for a monarch who, more than the rest of his dynasty or neighbouring rulers, invested his resources in elephants. (Cf. Nizam al-mulk in the Deccan in the early eighteenth century, who at one moment had with him 1,026 elephants, of which 225 were provided with armour and 'presumably were used in battle.'[29]).

Some of the estimates regarding the employment of elephants by the Mughal emperors in the sixteenth century appear to be very high. Akbar, when setting out against 'Ali-Quli Khan in Jaunpur, is said by his court historian to have 'chosen out 2,000 war-elephants to accompany the troops.'[30] But the same authority a few pages later says that he had only 'about 500 elephants' with him at the decisive engagement of

[27] *Ibid.*, p. 13.

[28] Farishta, *Ta'rikh-i Farishta*, Major-General John Briggs et al, eds, Bombay: Bombay Government College Press, 1831–32, vol. I, p. 563. Regarding elephants in the Deccan in the late fourteenth century, Farishta has some large and suspect figures. In the war of 1366 the Ray of Bijanagar (Vijayanagar) set out with a force of 30,000 horses, 900,000 foot and 3,000 elephants. See p. 550. Muhammad I Bahmani, according to a passage notorious in the controversy regarding the introduction of firearms into India, captured from the Ray 2,000 elephants, 300 gun-carriages and 700 Arab horses. See p. 552.

[29] Irvine, *The Army of the Indian Moghuls*, p. 180.

[30] Abu'l Fazl 'Allami, *Akbar-Nama*, Agha Ahmad 'Ali and Abdur Rahim, eds, Calcutta: Asiatic Society of Bengal, 1873–87, vol. II, p. 290; English Translation by Henry Beveridge, Calcutta: Asiatic Society of Bengal, 1897–1921, p. 246; Paul Horn, *Das Heer- and Kriegsgewesen der Grossmoghuls*, Leiden: E. J. Brill, 1894, p. 52.

this campaign.[31] Other figures of war-elephants in the campaigns of Akbar quoted by Horn are smaller than this;[32] while the statement by Badayuni that the rebel general Hemu had 1,500 elephants in his army appears improbable,[33] as does Farishta's assertion that the central Indian Queen Durgavati had the same number.[34]

After the end of the sixteenth century (and possibly for some decades before this), with more rapidly firing artillery and handguns the elephant, in Irvine's words, 'ceased even in the East to be of much value in the fighting line of battle.'[35] 'To the last some elephants protected by armour were brought into the battlefield. But their use was confined almost entirely to carrying generals or great nobles, and displaying their standards'.[36] Contemporary estimates of the number of elephants kept by the late sixteenth- and seventeenth-century Mughal emperors varied wildly, from 400 to 30,000:[37] but we may take as probably true for its day, around 1650 CE the evidence of that sober observer Tavernier, who states that he had inquired from the keeper of the royal elephants in the capital city (Shahjahanabad) how many elephants he had under his charge, and the latter had replied that he had 500 elephants 'of the household', of which 'only 80 or at most 90' were used for war.[38] Such *khassa* (royal) elephants, although for the most part not used for war by this time, would probably be animals of sufficient quality to train as war-elephants. In this period also there were probably considerable numbers of royal elephants stabled away from the court in provincial

[31] *Akbar-Nama*, vol. II, p. 293; Beveridge, p. 431.

[32] Horn, *Das Heer*, p. 52.

[33] Horn, *ibid.*; Badayuni, *Muntakhab al-tawarikh*, W. Nassau Lees et al, eds, Calcutta: Asiatic Society of Bengal, 1865, vol. II, p. 14. Badayuni also states that nearly 1,500 elephants were captured after the battle. See p. 16 of the same volume.

[34] Horn, *Das Heer*, p. 52; *Ta'rikh-i Farishta*, vol. I, p. 481: Badayuni in vol. II, p. 66 mentions that Durgavati gave her 700 elephants, while Abu'l Fazl in *Akbar-Nama*, vol. II, pp. 211, 214 states that 1,000 elephants were taken after her defeat and also that she had 1,000 'famous elephants' (*filan-i nami*).

[35] Irvine, *The Army of the Indian Moghuls*, p. 179.

[36] *Ibid.*, p. 175.

[37] Jean Baptiste Tavernier, *Travels in India,* trans. V. Ball, London: Macmillan & Co., 1889, vol. II, p. 224n, where varying estimates are quoted.

[38] *Ibid.*, p. 224.

establishments and in those pil-khanas in the countryside which have given their names to a number of north Indian villages. But a number of potential war-elephants does not seem to be implied which would be out of all proportion to the number available in the thirteenth- and fourteenth-century Delhi Sultanate.

For the total number of war-elephants in the pil-khana of the Delhi Sultans at the height of their power we possess two estimates. Sultan 'Ala al-din Khilji in the third year of his reign (1299 CE), according to the variant recension of Barani's history, had 'one thousand and five hundred' elephants in his pil-khana.[39] When rewriting his history in the final version, the same historian changed this figure to 'many elephants'[40] and it is therefore difficult to know if any reliance should be put upon it at all.

Al-'Umari provides an even larger figure of 3,000 elephants. Al-'Umari's descriptions of India are (apart from some traditional and legendary matter) based upon detailed questioning of returning travellers. In general his information must be taken as referring to the middle years of Muhammad bin Tughluq's reign, around 1340 CE. 'The Sultan of Delhi,' he writes, 'has 3,000 elephants of different kinds and sizes, and for the expenditure on their food the revenue of a large kingdom would hardly suffice. The largest variety require 40 *ratls* of rice and 60 of barley (*sha'ir*) daily, with 20 *ratls* of butter and half a load (*himl*) of hay. Beyond this there is the expenditure on their attendants. The *shihnat al-fila* is a great man with an *iqta* 'the size of a great land like "Iraq".'[41] These elephants were 'of different kinds' and therefore, even if al-'Umari's figure is not exaggerated, as in the case of Muhammad III Bahmani's elephants, only a portion of these would be war-elephants. If a quarter of them were, it would give Sultan Muhammad bin Tughluq a force of 750 war-elephants. In the following reign, when a decline in the total resources of the Delhi Sultanate had taken place, Feroz Shah Tughluq set out on major campaigns against Bengal and against Sind with 470 and 480 elephants respectively.[42] A maximum number of 750 to 1,000 war-elephants in the possession of the Delhi Sultans in the

[39] *BTFS-A,* folio 97A.
[40] *BTFS,* p. 262.
[41] *Masalik*, pp. 51–52.
[42] *ATFS,* pp. 144, 197.

period of their greatest power in the early fourteenth century appears not unlikely. Before the conquest of Gujarat and the plundering of the Hindu states of the Deccan the total number of war-elephants was probably not so high.[43] After 1350, in the third quarter of the fourteenth century the number of war-elephants in the pil-khana was probably maintained by the strenuous efforts of Feroz Shah Tughluq at a little below 500. In that monarch's old age and in the strife after his death the pil-khana evidently declined swiftly to the mere 120 beasts which the Sultan of Delhi could put into the field in 1398 CE.[44]

Accessions to the pil-khana, even of insignificant numbers of elephants, are frequently mentioned by historians of the Delhi Sultanate, from whose references it is possible to compile the following catalogue of entries:

(i) In 1192, 14 elephants were acquired with the conquest of Delhi, Kuhram and Ajmer.[45]

(ii) In 1194, 100 elephants were acquired from the expedition which defeated Jitacandra of Kanawj, according to one contemporary source;[46] other sources state 90 or 80 elephants.[47]

(iii) In 1225, Ghiyas ud-din 'Ivaz Khilji, established as ruler of Bengal, sent to Delhi 30 elephants as tribute.

(iv) In 1259, two elephants arrived from Lakhnauti (Bengal) as tribute.[48]

(v) In 1264, 60 (or 63) elephants were sent from Bengal to Delhi as tribute, evidently to conciliate the new Sultan, Balban.[49]

[43] Barani's reconsideration of the number of 1,500 elephants may have been prompted by the recollection that he was writing of the period before the Deccan expeditions of Malik Kafur.

[44] Muhammad b. Mansur b. Sa'id Mubarakshah, 'Fakhr-i Mudabbir', *Ta'rikh-i Fakhr al-din Mubarak Shah* (henceforth *TFM*), E. Denison Ross, ed., London: Royal Asiatic Society, 1927, pp. 80–81.

[45] *Ibid.*, p. 22.

[46] *Ibid.*, p. 23.

[47] Ibn al-Athir quoted in Henry Miers Elliot and John Dowson, *The History of India as Told by its Own Historians*, London: Trubner & Co., 1867–77, vol. II, p. 25: Hasan Nizami, *Taj al-Ma'athir* also quoted in the same volume, p. 233.

[48] Abu 'Umar Minhaj al-din b. Siraj al-din al-Juzjani, *Tabaqat-i Nasiri* (henceforth *TN*), W. Nassau Lees, ed., Calcutta: College Press, 1864, pp. 171, 226.

[49] *BTFS*, p. 53.

(vi) In 1281–82, an expedition despatched by Sultan Balban against Jajnagar (Orissa) killed four of the elephants of its ruler and captured 20. The Ray, entreating for peace, agreed to surrender 50 more elephants.[50]

(vii) In 1295 31, elephants were taken in the raid of the future Sultan 'Ala al-din upon Devgir, of which 30 fell to him on the day of his arrival at the city, which he surprised in the absence of the Ray.

(viii) In 1308–09, 17 elephants were taken by Malik Kafur, general of 'Ala ud-din, from his sack of Devagiri.[51]

(ix) In 1309–10, during the second Deccan expedition of Malik Kafur, three huge elephants belonging to the Ray of Arangal, which had been concealed at some distance from his fortress, were pursued and captured.[52] When the Ray entreated with the invaders 'hundreds' or 100 elephants were surrendered.[53]

(x) In 1310–12, during the third Deccan expedition of Malik Kafur, the Ray of Tilang (Arangal) despatched 23 elephants to join the Muslim army.[54] The Hoysala ruler at Dhorasamudra surrendered '36 huge elephants'.[55] One hundred and eight (or 120) elephants were captured from Vira Pandya at Kannur (Cannanore?).[56] Two hundred and fifty elephants were captured from the same

[50] Amir Khusrau Dehlavi, *I'jaz-i Khusravi*, Book V, pp. 9–11, Ms, ff. 385A–387A, Writer's Collection, Accession No. 34, Bodleian Library, Oxford, which reads *bist*, 'twenty' in place of *shast* (with *sin*) 'sixty' in the lithographed edition, as the number of elephants taken in battle. The information is suspect, as it is possible that this *fathnama* is a later imaginative composition of Amir Khusrau.

[51] *BTFS*, pp. 223, 228, 326. On the first occasion Barani writes *si u and*, 'upwards of thirty', and on the second *si u yak*, 'thirty one'. It is possible that one or other reading is corrupt.

[52] *KF*, p. 88.

[53] *KF*, p. 101 reads *sadha*, 'hundreds'. *Daval Rani Khizr Khan* (henceforth *DRKK*), Rashid Ahmad Ansari, ed., Aligarh: Madrasat-ul-um, 1917, pp. 69–70, twice says *sad*, 'a hundred'.

[54] *KF*, p. 120.

[55] *KF*, p. 138.

[56] *KF*, p. 153. Kundur is identified as Cannanore by S. K. Aiyangar in his introduction to Habib's translation of *KF*, p. xxxv.

ruling family at Brahmatpuri (Chidambaram?).[57] But only '2 or 3' elephants fell into the Muslim hands at Madurai in the same kingdom.[58] In all 512 elephants were entered into the Sultan's muster-roll by the end of the expedition.[59]

(xi) In 1312, on the return of Malik Kafur's army, according to Barani, 612 elephants were paraded before the Sultan among the spoils.[60] The figure is evidently misquoted from the 512 of Amir Khusrau's immediately contemporary account.[61]

(xii) At the end of 1312 20 more elephants reached Delhi as tribute from the Ray of Arangal.[62]

(xiii) In 1317–18 the Ray of Arangal, besieged again by Khusrau Khan surrendered more than 100 elephants.[63] An annual tribute of 100 elephants was imposed upon him.[64]

(xiv) In 1322 the unfortunate Ray of Arangal was attacked and carried away to Delhi by the future Sultan Muhammad bin Tughluq, with all his riches, including an unspecified number of elephants.[65]

(xv) In the same year this prince also marched on Jajnagar where he took 40 elephants which were destpatched to Delhi.[66]

(xvi) In 1324 Sultan Ghiyas ud-din Tughluq invaded Bengal and overcame both the rival Sultans of Bengal. Regarding the conquest of the second of these, Barani remarks: 'All the elephants of the country were sent to the royal pil-khana and the army acquired great spoil in the campaign.'[67] As Bengal was an area where war-elephants were trapped locally as well

[57] *KF*, p. 156. Brahmatpuri is identified as Chidambaram by S. K. Aiyangar, *South India and her Muhammadan Invaders*, Madras: Oxford University Press, 1921, pp. 99, 108–9.

[58] *KF*, p. 160.

[59] *KF*, p. 161.

[60] *BTFS*, p. 333.

[61] *KF*, p. 161.

[62] *BTFS*, p. 334.

[63] Amir Khusrau Dehlavi, *Nuh Sipihr*, Muhammad Wahid Mirza, ed., Calcutta: Oxford University Press, 1948, p. 120; *BTFS*, p. 398.

[64] *Nuh Sipihr*, p. 120.

[65] *BTFS*, p. 449.

[66] *Ibid.*, p. 450.

[67] *Ibid.*, p. 451.

as imported, the number taken to Delhi on this occasion may have been considerable.

For the reign of Muhammad bin Tughluq (1324–51), when the Delhi Sultanate was at the height of its splendour, no specific references survive concerning the capture or acquisition of elephants to the pil-khana, although Ibn Batuta alludes to the trade in elephants from Ceylon to mainland India.[68] As Muhammad's kingdom stretched to the extreme south of the Indian peninsula, there was no possibility of sensational accessions to the pil-khana by despoiling hitherto undespoiled Hindu princes; and throughout his reign the Sultan was mainly preoccupied with the suppression of his rebellious Muslim subjects—while his chronicler Barani is preoccupied with explaining the disasters which befell him. Barani's moralisings on the necessity of controlling Bengal, in order to procure elephants, probably indicate that this important source of supply was cut off by the rebellion of the province in the 1330s. But the great numbers of elephants acquired between 1309 and 1324 CE may have left a surplus in the pil-khana, where, in view of the longevity of the animals and their small casualties in battle, the need for replacements would not have been felt for many years. However the acquisition of elephants is known to have been a major preoccupation of Muhammad's successor, Sultan Feroz Shah Tughluq (1351–88):

(xvii) In 1354, Feroz Shah Tughluq made his first invasion of Bengal. Sultan Shams ud-din Ilyas drew up in battle before him with 50 elephants.[69] Three of these were slain and the remaining 47 captured.[70] Three more would appear to have been lost or given away, as only 44 were displayed in the victory celebrations at Delhi.[71]

(xviii) In 1359, five elephants arrived as tribute from Bengal; the small number incensed the Sultan.[72]

[68] Ibn Batuta, *Rihla*: *Travels* (selections), trans. H. A. R. Gibb, London: Broadway House, 1929, p. 77.

[69] *ATFS*, p. 114.

[70] *Ibid*., p. 118.

[71] *BTFS*, pp. 593–94.

[72] Yahya b. Ahmad b. 'Abd Allah al-Sirhindi, *Ta'rikh-i Mubarakshahi* (henceforth *TMB*), M. Hidayat Hosain, ed., Calcutta: Asiatic Society of Bengal, 1931, p. 128.

(xix) In the same year Feroz Shah Tughluq invaded Bengal for the second time. In the negotiations for peace the Sultan of Bengal sent 40 elephants to him[73] and in addition promised a yearly tribute of 40 elephants.[74]

(xx) Feroz Shah Tughluq then retired to Jaunpur and proceeded on an expedition into Jajnagar (Orissa), the principal purpose of which seems to have been to acquire elephants. The Ray of Jajnagar had fled from his capital, leaving an indomitable elephant standing loose in front of his throne-room: as this could not be captured; it had to be slain.[75] The Sultan then organised the hunt in which he captured eight wild elephants.[76] The Ray then entreated with Feroz and surrendered 20 elephants, or 18 according to another source, which adds that he was allowed to keep one elephant.[77] The Sultan, after a period when he was lost with his army in the Rajmahal hills, returned to Delhi with 73 new elephants, of which the provenance of five is unrecorded.[78] For the remaining 28 years of his reign Sultan Feroz Shah Tughluq is not recorded as having himself captured any more elephants though reference is made to his arrangements for purchasing elephants from overseas.[79]

(xxi) In 1370–71, Malik Raja Faruqi, in possession of Thalner in central India, is said to have captured five large and ten small elephants from a ruler called Raja Bharji or Biharji which were despatched to Sultan Feroz Shah Tughluq in Delhi.

This is the last recorded entry into the Delhi pil-khana before the debacle of 1398 CE but probably elephants were being purchased from overseas at a later date than this: Sirhindi, writing some 40 years after the event, maintains that a yearly tribute of elephants continued to be sent to Delhi from the Ray of Jajnagar and the ruler of Lakhnauti

[73] *Ibid.*, p. 161.

[74] *Ibid.*, pp. 161, 163.

[75] *Ibid.*, p. 166.

[76] *Ibid.*, pp. 168–69.

[77] *Ibid.*; *Sirat-i Ferozshahi*, quoted in J. M. Banerjee, *History of Firozshah Tughluq*, New Delhi: Munshiram Manoharlal Oriental Publishers, 1967, pp. 42, 185n.

[78] *ATFS*, pp. 172, 173.

[79] *Ibid.*, p. 486.

(Bengal) until 1394. In that year Khwaja-i-Jahan Sarwar, given the title of *Sultan-us-Sharq* (chief of the east) and despatched eastwards (with 20 elephants) to combat Hindu insurrections, established himself at Jaunpur. Thereafter these tributes of elephants were despatched to him.[80]

The Source of Supply

Elephants could come into the Sultan's pil-khana as plunder, as tribute from subordinate rulers or governors, by purchase from outside territories or trapped directly from a wild state. The relatively high figures, quoted above, of the contents of the Ghaznavid pil-khana suggest that by the thirteenth and fourteenth centuries the wild elephant may have grown less plentiful in northern India. But in the sixteenth century elephant trapping was still carried out extensively in areas to the south of the Gangetic plain. Kabir, who was probably living in the vicinity of Benaras around 1500 CE employed in his popular verses similes of the breaking in and training of elephants.[81] The emperor Babur, writing not long before 1530 CE mentions that the inhabitants of 30 or 40 villages in the district of Karra (modern Allahabad) gained a living as trappers of elephants; and that the wild elephant grew more plentiful towards the east.[82] The emperor Akbar (1556–1605) was connoisseur of elephants, and it was thought that the choicest in his possession came from Pannah,[83] (about which Watt remarks that it is 'a region where the elephant rarely, if at all, now exists').[84] However many other places in northern India are also mentioned from which Akbar procured some of his elephants.[85] His successor Jahangir, in a royal hunt in Gujarat (a territory which had been firmly under the control of the fourteenth-century Delhi Sultans)

[80] *Ta'rikh-i Farishta*, vol. II, p. 542; also see *TMB*, pp. 156–57.

[81] Kabir, *Dohavali*, Charlotte Vaudeville, ed., Pondicherry: Institut d'Indologie Francais, 1957, p. 58.

[82] *Babur-Nama*, trans. Annette Susannah Beveridge, London: Luzac & Co., 1922, p. 488.

[83] Abu'l Fazl 'Allami, *A'in-i Akbari*, trans. H. Blochmann and D. C. Phillot, Calcutta: Asiatic Society of Bengal, 1927, vol. I, pp. 129–30.

[84] Watt, *Dictionary of the Economic Products*, vol. III, p. 209.

[85] *A'in-i Akbari*, pp. 129–30.

captured 69 elephants.[86] But in the literature of the thirteenth- and fourteenth-century Delhi Sultanate, there is no reference to any local source of supply within the control of the Sultans, while the chroniclers painstakingly record the arrival of small numbers of elephants from distant lands.

The descriptions of a single elephant hunt of the Delhi Sultans survives, a hunt which took place during Feroz Shah Tughluq's raid upon Jajnagar. The eight large elephants (one being a female of great ferocity) were captured by the *kheddah* or enclosure method. 'Afif, who mentions a wooden palisade (*katgarh*), has not very clearly understood that the animals were driven into it. Tame elephants and vast numbers of beaters with loud musical instruments were used to contain the elephants. 'Afif states that when the captured animals had been sufficiently reduced by starvation, the *mahout*s climbed out on the branches of trees and dropped onto the backs of the elephants. This would, of course, be a very perilous action, if they had not previously persuaded them to accept from their hands the food which they otherwise lacked; but 'Afif does not mention this circumstance. However this hazardous method of taming was also practised in fifteenth-century Vijayanagar; and by the Mughal emperor Akbar in person.[87]

When elephants were wrested by the Delhi Sultans from other Indian rulers, it does not follow that they were native to that part of India. The elephants taken from the Ray of Jajnagar or from his dominions on three separate occasions are likely to have been locally trapped; but, as we know from a remark of Amir Khusrau, this was definitely not the case with those taken from the Ray of Arangal.[88] Even if they did not spend treasure with the lavish hand of the Delhi Sultans, the Hindu rulers of the Deccan had been devoting a portion of their resources to a medieval Indian armaments race, whose consequence we see in the immensely profitable Arab and Persian Gulf horse trade. The princes of

[86] Nur al-din Jahangir, *Tuzuk-i Jahangiri*, Henry Beveridge, ed., trans. Alexander Rogers, London: Royal Asiatic Society, 1909–14, vol. II, p. 12. Similarly in 1566, emperor Akbar is said to have captured 350 elephants in a single day, see Nizam al-din Ahmad, *Tabaqat-i Akbari*, Kanpur: Munshi Nawal Kishore, 1875, p. 265.

[87] *ATFS*, pp. 168–69; also see, S. H. Hodivala, *Studies in Indo-Muslim History*, Bombay: S. H. Hodivala, 1939–50, vol. I, p. 419.

[88] *KF*, p. 101.

India, including the Delhi Sultans, put their faith in elephants in warfare; besides which the actual possession of elephants was an important symbol of royalty and independent power. Idrisi, writing in Sicily in the twelfth century, had heard of the lust of the Kings of Hind to possess great and tall elephants, upon which they spent great sums of money, paying for them according to their height.[89] Cosmas Indicopleustes, the sixth-century Christian monk, writes of Kings 'of different places in India, who keep elephants, such as the Kings of Orrhotha, and the King of the Kalliana people, and the Kings of Sindu, of Sibor and of Male. One will have 600 elephants, another 500, and so on, some more and some less.'[90] The testimony of Megasthenes makes it clear that the acquisition of elephants from distant sources of supply was established in northern India by the third century before Christ.[91]

[89] Idrisi, *Nuzhat al-Mushtaq*, on India, S. Maqbul Ahmad, ed., Aligarh: Aligarh Muslim University, 1954, pp. 11, 79; English Translation by the same author, *India and the Neighbouring Territories*, Leiden: E. J. Brill, 1960, pp. 36, 71.

[90] *The Christian Topography of Cosmas Indicopleustes*, trans. J. W. McCrindle, London: Hakluyt Society, 1897, pp. 371–72.

[91] J. W. McCrindle, *Ancient India as described by Megasthenes and Arrian*, Bombay: Thacker; London: Trubner & Co., 1877, passim. The fragments which have been preserved of Megasthenes describe Candragupta Maurya at Pataliputra as the possessor of 8,000 or 9,000 elephants, pp. 139, 155–56. The rulers of other tribal groups are said to have possessed very much smaller numbers:—

The king of Automela	1,600 elephants (p. 147)
The king of the Andarai	1,000 elephants (p. 138)
The king of the Gangaridai	700 elephants (p. 155)
The king of the Kalingai	700 elephants (p. 136)
The king of the Pandai	500 elephants (p. 147)
The king of the Toluktai	400 elephants (pp. 137–38)
The king of the Charmai	'but 60' elephants (p. 147)

Before the rise of the Mauryas the Gangaridai are said to have possessed 4,000 elephants. See p. 34. The Varetatae or Suarataretae 'keep no elephants but trust entirely to their horse and foot'. See p. 146. References which imply a trade in elephants are to 5,000 councillors, each of whom furnished the state with an elephant. See p. 67; to the method of training elephants, pp. 90–91; to a 'half-wild' class of elephant hunters, pp. 136–37; to an elephant catcher unwilling to sell an animal to a king, pp. 118–19: and to the royal monopoly of elephants, p. 90. For the trade from Ceylon to India, see Malik Muhammad Jayasi, *Padmavat*, V. S. Agrawala, ed., Chirgaon (Jhansi): Sahitya Sadan, 1955,

It is obvious to Barani, writing in the middle of the fourteenth century, that the main source of war-elephants for the Delhi Sultans was Bengal. Down to recent times the thick jungles of eastern Bengal have been an important source of wild elephants. No figure of elephants arriving in Delhi from Bengal is recorded comparable to the magnificent 512 brought back from the Deccan by Malik Kafur, but there may have been a steady supply through the years which did not attract the attention of the chroniclers. Bengal in the thirteenth and early fourteenth century was sometimes under the direct control of the Delhi Sultans, when the *muqta*ʿ or governor was not expected to retain any elephants in his own control.[92] In other periods the Muslim rulers of Bengal were in a fairly amicable tributary relation, which had to be maintained by the occasional despatch of elephants to Delhi. Barani attributes a letter to Sultan Ghiyas ud-din Balban, addressed to his son Bughra Khan when he gave him the province of Bengal. The Bengal ruler should continue to send presents, offerings and trustworthy and well-disposed emissaries to Delhi so that the Sultan of Delhi should not consider an expedition against the Kingdom of Lakhnauti (Bengal) the most urgent of matters. From time to time he should send a certain number of elephants to Delhi, which would have the consequence that the Sultan of Delhi would not close the road for horses (from the north-west or the Arabian Sea) to reach Bengal.[93] There was in fact a *quid pro quo* in the relationship. When Feroz Shah Tughluq had, at the peace after his second invasion of Bengal in 1359, extracted 40 elephants from the Sultan of Bengal, he presented him with 500 fine horses.[94] This would hardly be a fair exchange if one reckons, as Barani did, a single elephant to be worth 500 horses: but then these were fine Arab and central Asian horses. On another occasion, when a strong and successful warrior, Sultan Balban, had usurped the throne of Delhi, prudence dictated the despatch by Tatar Khan from Bengal of a particularly large tribute of elephants.[95]

---

p. 391, stanza 385; also see *Padmavati*, trans. A. G. Shirreff, Calcutta: Royal Asiatic Society of Bengal, 1944, p. 232.

[92] *BTFS*, pp. 82–83; the case of the *muqta*ʿ Mughith al-din Tughril during the reign of Balban.

[93] *BTFS*, p. 196.

[94] *ATFS*, p. 159.

[95] The 60 (or 63) elephants which arrived in the year of Balban's accession, see *BTFS*, p. 53.

Many of the Bengal elephants may have come from beyond the frontiers of the territories firmly held by the the Sultans or Muslim governors of Bengal. From the territory of the Ray of Jajnagar Mughith al-din Tughril, Balban's governor of Bengal, was reported to have been taking elephants which had not been despatched to Delhi.[96] Other elephants were being brought into Bengal by sea, as we know from the complaint of Feroz Shah Tughluq recorded as made in 1355 or 1356 that the presents offered by Shams ud-din of Bengal were not adequate: in future they should include choice elephants of the sort acquired from sea-ports (*az banadir*).[97]

The 360 odd elephants taken by Malik Kafur from the Pandya Kingdom alone—'Ma'bar', the extreme south of the Indian peninsula—suggests a great plenty in that area. The 100 or more elephants surrendered the previous year at Arangal are also said to have been from Ma'bar: 'They are all "must" elephants of Ma'bar, not vegetable eating elephants of Bang (Bengal).'[98] In the extreme south of India, the final range of the Western Ghats as well as Coorg and parts of Mysore to the north are a habitat of wild elephants, which were extensively trapped there in the late nineteenth century. On the other hand we know from inscriptional evidence that the Pandya kingdom was in the mid-thirteenth century taking elephants from Ceylon[99] and there are many other references to this easy sea-borne trade across the narrow straits, of an earlier as well as of a later date.[100] A difficulty in identifying

[96] *BTFS*, pp. 82–83.

[97] *TMB*, p. 126.

[98] *KF*, p. 101.

[99] K. A. Nilakanta Sastri, *The Pandyan Kingdom*, London: Luzac & Co., 1929, pp. 161–62, 176; inscriptions of the reign of Jatavarman Sundara Pandya in the mid-thirteenth century at Tirupundurutti (no. 166 of 1894) and at Pudukottai (no. 366). The historical poem of the latter, which mentions a tribute of elephants from Karnataka and of elephants and pearls from Ceylon, is omitted in *A Chronological List of Inscriptions of Pudukottai*, Pudokottai: Sri Brihadamba State Press, 1929, p. 54.

[100] It is uncertain to what extent this trade was interrupted by the frequent and savagely conducted warfare between the medieval dynasties of South India and of Ceylon. Dr Padmanabhan informs me that the Telugu poem *Harivilasam*, written in the reign of Devaraya II of Vijayanagar (1422–46 CE), refers to a patron of the poet who was a 'shetty' or merchant supplying the Vijayanagar and Bahmani courts with elephants from Ceylon; and in the fifteenth century

Malik Kafur's Ma'bari elephants as of Sinhalese origin is the statement that the elephants were young and still possessed both their tusks.[101] The male of the Sinhalese elephant is generally tuskless, and as it is closely related in the same species to the Indian elephant, this is a minor genetic mystery: it has been suggested that the tusked Sinhalese male elephant has been bred out by the particularly ruthless demand for ivory in that country.[102] We cannot therefore assume that in the fourteenth century male Sinhalese elephants were tuskless.

There is no doubt of the antiquity of the trade in Sinhalese elephants to much more distant areas of the Indian sub-continent. Megasthenes refers to their presence at the third-century BCE Mauryan court at Pataliputra in Bihar.[103] Cosmas, writing in the sixth century CE mentions the good price which the King of Ceylon (Seiladibe) gave for the elephants, at a fixed rate according to their height.[104] He mentions this in a passage concerning both the import of horses into Ceylon and the holdings of elephants of distant Indian monarchs; their export as a royal monopoly is therefore implied. The detail of Sinhalese elephants being sold according to their height is mentioned by many travellers, among them 'Abd al-Razzaq writing in the mid-fifteenth century[105]

---

the trade to south India seems to have been well established. See 'Abd al-Razzaq Samarqandi, *Matla' al-Sa'dayn*, M. Shafi', ed., Lahore: Kitabkhana-i Gilani, 1941–49, vol. II, p. 805; *The Suma Oriental of Tome Pires*, Armando Cortesao, trans. and ed. London: The Hakluyt Society, vol. I, p. 86; A. Nikitin translated in R. H. Major, *India in the XVth Century*, London: The Hakluyt Society, 1857. Pires refers to an annual tribute of 40 elephants from Ceylon to Quilon, interrupted by the Portuguese (p. 80). For the sixteenth- and seventeenth-century export of elephants from Ceylon to India see Linschoten, *Travels*, trans. Arthur Cole Burnell amd Paul Anton Tiele, London: The Hakluyt Society, 1885, vol. I, p. 81; also Paulus Edward Pieris, *Ceylon: The Portuguese Era*, Colombo: Colombo Apothecaries Company Limited, 1914, vol. II, pp. 66–68; S. Arasaratnam, 'Dutch Commercial Policy in Ceylon and its effects on the Indo-Ceylon Trade, 1690–1750', *Indian Economic and Social History Review*, 4: 2, June 1967, p. 109.

[101] *KF*, p. 101.

[102] Tavernier, *Travels in India*, vol. I, p. 222n.

[103] McCrindle, *Ancient India as described by Megasthenes*, p. 170.

[104] *The Christian Topography*, pp. 371–72.

[105] 'Abd al-Razzaq, vol. II, p. 805; Pires, vol. I, p. 86; A. Nikitin, p. 20.

and Pires some 70 years later.[106] The Sinhalese elephant was smaller than the greatest of Indian elephants but was valued for its reputed greater sagacity and bravery in battle.[107]

Sinhalese elephants were probably to be found in the royal elephant stables of north Indian princes in the period before the establishment of the Delhi Sultanate. 'Utbi records that 'the Thanesar expedition of 405/1014–15 was provoked by Mahmud (of Ghazna)'s desire to get some of the special breed of Sailamani elephants excellent in war...'[108] Hodivala in his commentary on this passage suggests that the elephants probably were not Sinhalese at all, but must have been wild elephants of the sub-Himalayan range near Thanesar. Possibly, he suggests, like the Sinhalese elephants these may be tuskless, and for this reason the name may have been transferred to them.[109] But even if the elephants of Ceylon were tuskless in the early Middle Ages, we have no other references to the bravery and desirability of these sub-Himalayan elephants. The early sixteenth-century poet Jayasi, writing in a rural environment in the Gangetic plain, constantly refers to the excellence of Sinhalese elephants. 'Thousands of strings of horses went with them and a hundred strings of Singhali elephants.'[110] 'Where are my horses, gallant and strong? Where are my Singhali elephants?'[111]

Barbosa, writing before 1518 CE twice refers to the transport of Sinhalese elephants to Gujarat. The Sultan of Gujarat is 'a lord of horses and elephants in great numbers. The elephants come from Ceilam and Malabar to be sold there.'[112] I. H. Qureshi who noted these references argued that it would have been too expensive to import Sinhalese

[106] Pires, vol. I, p. 86.

[107] Watt, *Dictionary of the Economic Products*, vol. III; Linschoten, *Travels*; Vladimir Minorsky, *Marwazi on China, the Turks and India*, London: Royal Asiatic Society, 1942, pp. 46–47; McCrindle, *Ancient India as described by Megasthenes*, p. 170 appears to have wrongly believed that the Sinhalese elephant was larger than the Indian, but he also alludes to its reputation for superior intelligence.

[108] Bosworth, *The Ghaznavids*, p. 116.

[109] Hodivala, *Studies in Indo-Muslim History*, vol. I, pp. 144–45.

[110] Malik Muhammad Jayasi, *Padmavat*, p. 391, stanza 385: *Padmavati*, Shirreff, p. 232.

[111] *Ibid.*, p. 409, stanza 404; *Padmavati*, p. 242.

[112] *The Book of Duarte Barbosa*, trans. M. Longworth Dames, London: The Hakluyt Society, 1918–21, vol. I, pp. 117–18; vol. II, p. 113.

elephants to thirteenth- and fourteenth-century Delhi:[113] but medieval Indian princes did not hesitate to spend upon elephants and horses.

Two contemporary references appear to allude to the traffic of Sinhalese elephants to the capital city of Delhi, probably through a port in Gujarat. Ibn Batuta, describing the elephants in Ceylon, says that some of them are transported *ila hazrati mulk-ul-Hind*.[114] In this period *mulk-ul Hind* may be equated with the Delhi kingdom, while *hazrat*, in the fourteenth-century Indo-Persian usage with which Ibn Batuta after several years of residence was familiar, refers to the capital city of the Sultan. The Sinhalese elephants would have been arriving at the pil-khana of Muhammad bin Tughluq.

The second reference to this traffic is by 'Afif, and must allude to a fairly late period in the reign of Feroz Shah Tughluq—certainly after 1369. A royal *farman* (royal order bearing the seal of the emperor) had been issued to the effect that if any merchant should bring elephants from the *jaza'ir-i rod-i nil*, the price of any elephant which had perished on journey would be paid by the treasury. Acting on this farman, Shams ud-din Aburja—a minister of the Sultan for whom 'Afif had a great dislike—had fraudulently collected the price for several elephants from the *diwan-i-wizarat*.[115] *Jaza'ir-i rod-i nil*, 'the islands of the river Nile' or 'the islands of the blue water' is an unfortunate but not really misleading phrase: for 'Afif maintained through the hundreds of pages of his history his own remarkably crude style of *saj*' (rhythmical prose with internal rhymes) in which *pil* or *fil*, 'elephant', usually brings *nil* to mind.[116] Ceylon is the nearest and most likely source for a direct maritime trade in elephants to the ports of Gujarat, still at this time in the possession of the Delhi Sultans.

It is likely that Ceylon was in this period not the only source of elephants imported by sea into India. Barbosa, writing before 1518 CE referred to the export of elephants from the King of Pegu (lower Burma, near Rangoon), 'which they sell in many lands, but most of them in the kingdoms of Narsyngua (*i.e.*, Vijayanagar), Malabar and Cambaya (*i.e.*, Gujarat).'[117]

[113] Ishtiyaq Husain Qureshi, *The Administration of the Sultanate of Delhi*, fourth edition, Karachi: Pakistan Historical Society, 1958, p. 143.

[114] Ibn Batuta, *Rihla*: *Travels*, p. 77.

[115] *ATFS*, p. 486.

[116] cf. *ATFS*, pp. 197, 428–29.

[117] *The Book of Duarte Barbosa*, vol. II, pp. 154–56. Barbosa refers to the export of elephants from Champa to 'many lands' (these would probably

In the late sixteenth century very large numbers of elephants were trapped in Pegu, the realm of 'the Lord of the White Elephant'.[118] If Sinhalese sources are to be believed, in the mid-twelfth century not only were single elephants offered in tribute to Parakramabahu, King of Ceylon, whenever Sinhalese ships called at the Burmese coast, but Sinhalese merchants also purchased Burmese elephants to carry abroad (to Ceylon?)[119] There are therefore grounds for believing that a sea-borne trade in elephants across the Bay of Bengal from lower Burma was well established before the fourteenth century. When Feroz Shah Tughluq, in 1355 or 1356, complained that the Bengal Sultan was not sending him choice elephants 'from the sea-ports',[120] the elephants to which he is referring are likely to have come to the ports of Bengal from the delta of the Irrawaddy rather than from Ceylon.

The Role of the Pil-khana in the Downfall of the Delhi Sultanate

Amir Timur's supposedly irresistible role as a world-conqueror, as well as the scanty and contradictory chronicles of late fourteenth-century Delhi, have diverted attention from the rapid and disastrous breakdown of the Delhi Sultanate at the centre in the years from 1388 to 1398. This decade witnessed the dissipation of the greater part of the resources at the command of the Delhi Sultans in their capital city and the repeated failure to establish the authority of a single Sultan or a single mayor of the palace. In the struggle for power, possession of the pil-khana was one of the most

include the Indian subcontinent). See p. 209 of the same volume. In the seventeenth century elephants were being shipped in considerable quantities from Siam and also probably from Indo-China to the Indian market.

[118] See the narrative of Ralph Fitch in *Early Travels in India, 1583–1619*, William Foster, ed., London: Oxford University Press, 1921, p. 32; see also Moti Chandra, 'The White Elephant', *Lalit Kala: Journal of Oriental Art chiefly Indian*, nos. 1–2, 1955–56, pp. 96–97.

[119] Cyril Wace Nicholas, *A Concise History of Ceylon*, Colombo: Ceylon University Press, 1961, pp. 325–26.

[120] *TMB*, p. 126. In the seventeenth century, elephants were being shipped in considerable numbers from Siam and probably also from Indo-China to the Indian market.

important assets. In all the complicated manoeuvres of the decade the rule seems to hold good that he who held the elephants of the pil-khana and the loyalty of their attendants, could not be defeated in the open field by his opponent, however strong the support which the latter commanded.

The configuration of the capital city of Delhi in the late fourteenth century favoured a polarisation of politics into two main opposing factions, with their respective bases in the palace-precinct and garden city of Ferozabad and the older highly fortified triple city of *Dehli-i Kuhna*, *Siri* and *Jahan-Panah* some miles to the south. The number of defensible strongholds in the latter favoured a further division of power amid warring factions, which occurred in the three years before 1398. Civil war broke out in 1387 when Sultan Muhammad, second and eldest surviving son of Feroz Shah Tughluq, overthrew and slew the *wazir* Khan-i Jahan, and the Ferozshahi slaves rallied against him at Ferozabad around the persons of the aged Sultan and his grandson Tughluq Shah. When Sultan Muhammad had driven his opponents to the doors of the palace precinct (*kushk-i khass*), his aged and senile father was put upon his mount of state, brought out and displayed. At this, according to one source, the army and the elephant-drivers (*lashkar u pilbanan*) deserted Sultan Muhammad for their old master.[121] Sultan Muhammad retreated towards Delhi (the old cities), where the *Shihna* of the city closed the gate in his face. Routed, he fled with a small body of cavalry into the Punjab hills. In Ferozabad Tughluq Shah was installed upon the throne, with his supporters in possession of such of the *pa'egah* (royal stables) as was stationed in Delhi, and of the pil-khana.

Sultan Muhammad was the most able and persistent of the later princes of his house. He enjoyed extensive support among the free urban population, the great *iqta'*-holders and the lower military commanders.[122] He was able to muster 50,000 horses before two of his marches upon Delhi.[123] Yet he was repeatedly unable to gain a victory on the battlefield; and one reason for this may be sought in his opponents' possession of the pil-khana and employment of elephants in battle against him.[124] By contrast, his opponents were hindered in their

[121] *TMH*, folio 416A; *TMB*, p. 139.

[122] *TMH*, folio 421B; *TMB*, pp. 145, 148.

[123] *TMB*, pp. 146–47.

[124] *TMH*, folios 421B, 422A. These folios also describe Sultan Muhammad's battle with Abu Bakr Shah outside the walls of Delhi; that Abu Bakr takes

contest with him by feeling unable to leave the capital city unguarded for any length of time, probably in view of the sentiment of the free population in favour of Sultan Muhammad.[125]

After the murder of Tughluq Shah in a palace revolution and the flight of his successor Abu Bakr Shah to escape a similar fate, Sultan Muhammad accepted the invitation of his erstwhile enemies, the Ferozshahi slaves and the other amirs at Ferozabad to resume the throne there.[126] When he reached the banks of the Jamuna, all the *malik*s and amirs, with the entire pil-khana and pa'egah as well as the keys of the city-gates, came to wait upon him.

After his re-ascension of the throne and the distribution of offices and largesse, Sultan Muhammad was mainly preoccupied with how to wrest *de facto* control of the pil-khana and the pa'egah from the hands of the slave amirs. Both pil-khana and pa'egah were in Ferozabad. (The pil-khana was probably situated in the great rectangular walled enclosure at the southern side of the palace-precinct now known as Feroz Shah Kotla). Sultan Muhammad left Ferozabad for the older cities of Delhi, to visit the tombs of Sultans and Shaykhs, and began to show interest in the repair of the *kushk-i hazar-sutun* (the palace of the thousand pillars), a building with strong historical associations. As most of the slave amirs had households in the garden city portion of Ferozabad, they departed to them during the Sultan's expedition, and only a few remained on guard at the pil-khana and the pa'egah. Sultan Muhammad made a surprise attack upon these and gained possession of the elephants and horses. He then gave the Ferozshahi slaves three days to depart from Delhi. Many of them went to Mewat, to join Abu Bakr Shah, the Sultan who had recently fled from them. Those who remained in Delhi were hunted out and killed after three days. The elephants of which Sultan Muhammad gained control were shortly afterwards brought to bear upon the doors of a fortress some miles from Delhi, where his rival Abu Bakr Shah had taken refuge: and with the surrender of this prince Sultan Muhammad

'all his elephants, horses, Mughals and servants' and attempts to surprise Muhammad at Jatesar, but has to retreat when Muhammad makes for Delhi. In *TMB*, p. 138, there is a description of Abu Bakr defeating Humayun, son of Muhammad, when another large force of cavalry was raised against him.

125 *TMB*, p. 148.

126 *TMH*, folio 423A; *TMB*, p. 149.

enjoyed undisturbed possession of the throne for nearly three years until his death.[127]

In 1393, when the throne passed to Sultan Mahmud, son of Sultan Muhammad, the pil-khana again became a counter in the struggle for power in the capital city. The principal offices of state were divided among the great officers (*bandgan-i kibar*) of the Sultan's father, who very quickly fell out among themselves. Khwaja-i Jahan Sarwar departed to quell Hindu insurrections in the east, taking with him 20 elephants of the pil-khana; with his establishment at Jaunpur, the tribute of elephants from Jajnagar and Bengal is said to have ceased to reach Delhi.[128] An expedition of the Sultan with his wazir Sa'adat Khan to quell the Hindus of Bayana led to the establishment of a triumvirate within the old triple city of Delhi, consisting of Muqarrab Khan, probably a close relation of the Sultan, Mallu (Iqbal) Khan and Bahadur Nahar (or Bahadur Khan), a tribal chieftain from the neighbouring Mewat. Sultan Mahmud remained in the palace fortress of Ferozabad, under the control of Sa'adat Khan and Tatar Khan.[129] This pair brought out Sultan Mahmud 'with all the elephants and *hasham* (army attendants)' to besiege the triumvirate in Delhi. They were encamped at Hauz-i Khas outside the city walls, where the Sultan managed to flee from his tent to join their opponents, taking refuge with Muqarrab Khan in *Jahan-Panah*. 'Most of the population' of the city came out in a sortie to do battle for the newly arrived Sultan: but the other side still possessed the elephants and the rest of the hasham. The citizens were routed and left 1,000 horses and 'foot-archers' (*dhanuk*) dead on the field.

As Sa'adat Khan and Tatar Khan had lost control of the person of the Sultan, they put up a cousin of his as his substitute in Ferozabad, with the title of Nusrat Shah. Our source remarks that they kept Nusrat Shah as a figurehead and ruled according to their own counsels. In the end a difference arose between Sa'adat Khan and Tatar Khan, and the latter 'proved the stronger'. Sa'adat Khan fled with some of his supporters from the palace to the walled city, where he was received with honour but not long afterwards killed by Sultan Mahmud's

[127] *TMH*, folios 424B, 426A; *TMB*, pp. 150–51.

[128] *TMH*, folio 432B; *TMB*, p. 156.

[129] *TMH*, *Ibid.*; *TMB*, p. 158. Tatar Khan's alliance with Sa'adat Khan is not mentioned by *TMB* which speaks of him being raised to the *wizarat* and honoured with this title at a later date by Nusrat Shah, p. 160.

mentors.[130] Sirhindi, who up to this point has omitted to notice the role of Tatar Khan, gives a rather different account of Sa'adat Khan's flight. Some time after Nusrat Shah's elevation to the throne by Sa'adat Khan, he states, some of the Ferozshahi slaves[131] and some of the elephant keepers (*pilbanan*) made friends with the Sultan. They mounted Nusrat Shah upon an elephant and everyone began to rally round him. Sa'adat Khan saw that resistance was hopeless, and fled by the *haram* door to enter the city of Delhi.[132]

Fighting went on continuously between the supporters of Sultan Mahmud and Sultan Nusrat Shah in the few miles between their respective bases—every day through three years according to Sirhindi.[133] After the departure and death of Sa'adat Khan, Nusrat Shah began to be frightened of or dissatisfied with his remaining mentor, Tatar Khan. He would appear to have seized his opportunity when Tatar Khan, accompanied by 10 elephants and some of the royal slaves, left Delhi to combat Sarang Khan, who had control of Multan and was advancing towards the capital.[134] (Sarang Khan was the brother of Mallu Khan, but does not appear to have been acting in concert with him). Mallu Khan had been sending messages and making overtures to Nusrat Shah; Bihamad Khani describes Mallu Khan as a crafty man and Nusrat Shah as lacking in sense.[135] A meeting took place between the two on the neutral ground of the shrine of Qutb ud-din Bakhtiyar Kaki, outside the walls of *Dehli-i Kuhna*.[136] Mallu Khan promised to enthrone Nusrat Shah in the *kushk-i hazar sutun*, ancient throne-room of the Delhi Sultans in *Jahan-Panah*, the middle portion of the triple city, and to expel Sultan Mahmud.[137] Nusrat Shah therefore 'hastily and inauspiciously' departed from the palace of Ferozabad with all his elephants and military following. When he approached the city walls, Mallu Khan came out to receive him with 'some thousands' of horses, and he was installed, as promised, in the

[130] *TMH*, folios 432B–433B; *TMB*, pp. 158–59.

[131] (*sic*) but *TMB*, pp. 147, 150, has already described the massacre and almost total dispossession of the Firozshahi slaves in the provinces and in the capital.

[132] *TMB*, pp. 159–60.

[133] *TMH*, folio 433B; *TMB*, pp. 160–61.

[134] *TMB*, p. 161.

[135] *TMH*, folio 433B.

[136] *TMB*, p. 163.

[137] TMH, folio 433B.

kushk-i hazar sutun.[138] Sultan Mahmud and his kinsman Muqarrab Khan seem to have been obliged to retreat from this palace in *Jahan-Panah*, into the adjoining city to the south-west, *Dehli-i Kuhna*, where the Mewati chieftain Bahadur Nahar was already in residence. Within the triple walled city there were now lodged two Sultans and three would-be mayors of the palace. Mallu Khan openly supported Nusrat Shah while secretly entering into correspondence with Sultan Mahmud and Muqarrab Khan. 'As he did not think him worthy to rule,' Bihamad Khani observes, 'he took into his own keeping all his elephants.' Nusrat Shah then fled from the city 'in some direction'.[139] Sirhindi says that Nusrat Shah with his followers escaped from the city, but in the flight all the 'celebrated elephants' (*pilan-i namdar*) fell into Mallu Khan's hands. Nusrat Shah went again into Ferozabad, but after a while departed with all his following to join Tatar Khan,[140] distaste for whose tutelage had led him to negotiate with Mallu Khan: Mallu Khan by the capture of what was left of the pil-khana had himself, in Bihamad Khani's words, become 'the possessor of power and pomp.'[141] Of the other members of the triumvirate, Bahadur Nahar evidently decided that the immediate prospects in Delhi were poor and went off to his own tribal territories. Mallu Khan had gained possession of Ferozabad, but himself continued to reside at *Siri*, the north-eastern unit of the triple-walled city of Delhi, where he was installed before he lured Nusrat Shah to his downfall.[142] Sultan Mahmud and Muqarrab Khan were obliged to entreat with Mallu Khan. Sultan Mahmud once again moved from *Dehli-i Kuhna* into the *kushk-i hazar sutun* in *Jahan-Panah*, while Muqarrab Khan also took up residence close to this palace; but Mallu Khan 'with the prestige of all the elephants' kept the gates of the Sultan's residence closed. After suitably prolonged demonstrations of friendliness, Mallu Khan surrounded Muqarrab Khan's residence and brought him to a miserable end.[143] Mallu

[138] *Ibid.*; *TMB*, p. 163.

[139] *TMH*, folio 434A.

[140] *TMB*, p. 163.

[141] *TMH*, folio 434A; *TMB*, pp. 163–64.

[142] *TMB*, pp. 163–64, where it is stated that there were two months of fighting every day between Mallu Khan and Muqarrab Khan, presumably within the walls of the triple city.

[143] *TMH*, folio 434A; *TMB*, p. 164 adds that after guaranteeing his life he killed him.

Khan, with Sultan Mahmud in his power, then led out a force against Tatar Khan, Nusrat Shah's former wazir, at large to the north of Delhi. He defeated him and recaptured the (10) elephants of the pil-khana in his charge, which Tatar Khan had preferred to leave inside the walled town of Panipat while himself making a raid upon Delhi.[144] Now for a brief period Mallu Khan was undisputed mayor of the palace, with the only surviving Sultan and the entire pil-khana under his control.

The end of the pil-khana came very shortly afterwards, in December 1398 when the armies of Amir Timur of Samarqand drew up for battle outside Delhi. 'On the other side,' Timur's chronicler Nizam ud-din Shami remarks, 'was Malikzada Sultan Mahmud with Mallu Khan and other leaders and commanders of the land of Hind, with 10,000 horse and 20,000 fully armed foot[145] and 120 war-elephants, surging like the ocean and trumpeting like thunder clouds, armoured and with structures placed upon their backs.' Beside the elephants stood *ra'd-andazan* (throwers of explosive grenades) and *takhsh-afkanan* (either crossbowmen or rocketmen) while on top of each elephant there were several archers.

In spite of their long record of victories, the army of Timur showed some alarm at the sight of the elephants. Timur, observing this, ordered a certain Maulana in his train to spread the prayer-carpet and offered prayers to the Almighty for victory.[146] Practical preparations against the elephants included the erection of palisades and the digging of a trench in front of the army. In front of these, buffaloes were tied together, and

[144] *TMB*, p. 164.

[145] Forty thousand, according to a slightly later authority; see Sharaf al-Din 'Ali Yazdi, *Zafar-Nama*, Muhammad 'Abbasi, ed., Tehran: Amir Kabir, 1957, vol. II, p. 77; the figure of 12,000 horses, found in the translation of Yazdi in Elliot and Dowson, *The History of India*, vol. III, p. 498, appears erroneous, possibly from the reading of a corrupt manuscript. The information regarding the battle in both the *Zafar-Namas* appears to be derived from Ghiyath al-din 'Ali Yazdi's *Roznama-i ghazawat-i Hindostan*, of which the unique manuscript is in Tashkent. See Charles Ambrose Storey, *A Bio-bibliographical Survey of Persian Literature*, London: Luzac & Co., 1927, p. 278, no. 353. The microfilm of this manuscript at the India Office Library, London is unsatisfactory, but there appears to be no additional information in its florid account of the battle which is not reproduced by either Nizam al-din Shami or Sharaf al-din 'Ali Yazdi.

[146] Nizam al-Din Shami, *Zafar-Nama*, Felix Tauer, ed., Prague: Oriental Institute, 1937, vol. I, pp. 189–90.

smiths were ordered to manufacture caltrops to be distributed to the foot-soldiers to throw in the path of the elephants.[147] An initial attack on Timur's right by the Delhi forces was unsuccessful, being itself attacked in the rear by Timur's vanguard. As on other occasions, the Delhi elephants were mostly in the centre with the Sultan.[148] They advanced in good order but the warriors of Timur, in Shami's words, 'struck at the rank of the elephants and turned the elephant drivers upside down.'[149] The young prince Khalil Sultan, after killing the driver of an elephant, led it off to Timur 'as a husbandman drives a buffalo.'[150] From the Delhi side Sirhindi says that some elephants fell into the hands of the enemy in the battle, but Mallu Khan led back most of them by stratagems and hard driving inside the city walls. The Indian forces retreated to the walled city, where great numbers were crushed trying to enter the gates.[151] During the night, with Timur encamped outside, Mallu Khan and Sultan Mahmud secretly fled away, leaving the inhabitants of the cities of Delhi, whose great fortifications had never fallen to siege, to make their peace with Timur, only to be subsequently plundered, and enslaved. Yazdi states that among the plunder 120 elephants were paraded before Timur and sent off to Samarqand and other places within his dominions.[152] This is the same number which Shami mentioned in the battle-line; it suggests that, as on other occasions, few of the elephants were killed or seriously injured in battle. A curious detail is added by another early fifteenth-century Timurid chronicler. The elephants were laden not only with treasure but also with carved stone to adorn the great mosque of Samarqand with marble (*sang-i rukham*).[153]

This may be one of the causes of the puzzling and almost complete absence of elaborate stone carving on later fourteenth-century monuments in Delhi. There is no sign of Indian carved stone or marble upon the great mosque of Samarqand and it may be presumed that the elephants and this portion of their burden were mislaid or perished

[147] Yazdi, vol. II, p. 78.

[148] *Ibid.*, p. 83.

[149] Shami, vol. I, p. 190.

[150] Yazdi, vol. II, p. 84.

[151] *TMB*, p. 166.

[152] Yazdi, vol. II, p. 91.

[153] Mu'in al-din Natanzi, *Muntakhab al-tawarikh-i mu'ini*, Jean Aubin, ed., Tehran: Khayyam, 1957, p. 372.

upon the journey. Possibly, like the two parrots presented to Timur by Bahadur Nahar which could remember conversational phrases from the court of Sultan Ghiyas ud-din Tughluq three-quarters of a century before,[154] many of the elephants had grown old in the Delhi pil-khana.

The decline in the number of war-elephants from 470 early in Feroz Shah Tughluq's reign to a mere 120, which is more than matched by the decline in the number of horses from 80,000 or 90,000 to 10,000, shows how greatly the resources of the Delhi Sultans had diminished. But even 120 elephants were enough to cause concern in the nearly invincible armies of Timur and to maintain the semblance of allegiance to the Sultan's authority among the Muslim fief-holders of northern India. In Sirhindi's descriptions of the next few years, Mallu Khan and other noblemen marched upon the capital with four or 10 elephants[155] and the capture of two elephants in battle was noteworthy.[156] It is therefore possible to see the loss even of the reduced pil-khana of 1398 as what, more than anything else, placed the holders of the capital city of Delhi on terms of mere equality with those who now held power in Gujarat, Malwa and Jaunpur. The north Indian Sultanates had to build up their pil-khanas from scratch. To judge from the information of the early Portuguese sources, a balance of power was maintained in this renewed but traditional Indian armaments race.[157]

[154] Yazdi, vol. II, p. 99.

[155] *TMB*, pp. 167, 168.

[156] *TMB*, p. 169.

[157] *The Book of Duarte Barbosa*, vol. II, pp. 113, 154–56 (imports from Ceylon). Barbosa estimated that the King of Narsingua (Vijayanagar) possessed 900 elephants (vol. I, pp. 209–10); the King of the Cambaia (Gujarat) 400 or 500 (vol. I, p. 118). The estimates in Pires, vol. I are probably too modest; it alleges that the King of Gujarat possessed '300 elephants, about 100 of which are fighting elephants' (pp. 40–41); the King of Deccan 'about 50 elephants' (p. 52); the King of Vijayanagar 500 elephants (p. 64); the King of Delhi 'a very large number' (p. 90). Babur estimates that Sultan Ibrahim Lodi of Delhi and his great officers brought about 1,000 elephants to the battle of Panipat in 1525 CE. See *Babur-Nama*, p. 463. For a probably exaggerated estimate of the elephants of the Sultan of Jaunpur in the fifteenth century, see footnote 28 of this chapter. We have no estimate of the total holding of the Sultans of Malwa; 23 elephants were captured from the Malwa army by the Bahmani forces at the battle of Kherla in 1468 CE. For details see, Mahmud Gavan, *Riyaz al-Insha*, Husain Chand and Ghulam Yazdani, ed., Hyderabad: S. C. Government Central Press, 1948, p. 85.

The Imperial *Cheetahs* in Akbar's *Shikarkhana**

Seen through the Paintings of his Time

Asok Kumar Das

With its thick tropical jungles, vast stretches of grasslands, wooded hills, and difficult terrains, India sheltered much game and wildlife. Large tracts of land were earmarked as royal preserves where no one but the king was permitted to hunt, because hunting was a royal sport. From the hilly tracts of Ferghana in Uzbekistan the Mughals came to India, built a vast empire and amongst other things took great delight in this privileged activity. It relieved the monotony of court life and arduous military expeditions.

Members of the royal family, including the *zenana*, the women, the important courtiers and retainers, all participated in these hunts. It is said that emperor Jahangir once continuously spent three months and six days in the vicinity of Lahore, hunting. Akbar, too, was drawn to this sport and indulged in elaborate hunting exercises whenever he got the time. And lion hunting was a royal privilege everywhere. A special department called the *shikarkhana* was part of the palace administration and was under the charge of the chief huntsman, the *Mir Shikar*, who catered for and organised these elaborate royal hunts. Later, his position was re-designated the *Qarawal Begi*.

There were many specialised hunts tailored to the constraints of the area and its ecology: the *qamargah* or ring hunt, *shakhband* or where the area was fenced off with the branches and trunks of trees. They hunted with nets, with decoy animals, with hunting dogs and with trained cheetahs. For big game, kings often used muskets and handguns and at other times carried swords, javelins, bows and arrows.

* Asok Kumar Das, 'The Imperial Cheetahs in Akbar's Shikarkhana, seen through the paintings of his time', *The India Magazine*, volume 5, issue number 9, 1985, pp. 28–39.

The Mughals were keen on falconry and had a chief falconer, who looked after and nurtured their large collection of prized game falcons. Fishing and crocodile shooting were popular and Jahangir mentions in his memoirs how he liked the taste of a freshly caught *rohu* (a variety of fish). Elephants were most important in the royal paraphernalia and indispensable in war and during peace time, therefore, elaborate arrangements were made to trap these enormous mammals in their natural habitat across the country. Both Akbar and Jahangir took an active part in *khedda*s, or elephant-trapping operations, in the jungles of Dohad in north east Gujarat.

Deer hunting, using trained cheetahs, was a Mughal specialisation. The royal hunting grounds of present day Punjab, Haryana, Delhi, Rajasthan, south west Uttar Pradesh, western Madhya Pradesh and north Gujarat abounded in species of deer and the Mughals used the trained cheetah to get the prey. Cheetah hunting was apparently unknown to the Mughals before 1555 since Babur does not mention this sport in his detailed memoirs, the *Waqi'at-i-Babari*, and none of Humayun's biographers recorded anything about the cheetah. The first mention is by Akbar's biographer and principal councillor, Abu'l Fazl, in his *Akbarnama*, in the portion that describes Humayun's triumphant return to India in 1555.

In the later part of that year Akbar, then a young boy of 13, was introduced to this sleek and beautiful animal by Wali Beg, father of a leading courtier, Khan Jahan. Wali Beg got a cheetah, named Fateh Baz, from the Afghans at the battle of Maciwara. The cheetah came with the able keeper, Dundu Khan, who was given the title Fateh Khan when he was absorbed by Akbar as a prime huntsman. Viewing this sport and receiving his first cheetah, Akbar was infatuated by this sport and the animal. A large number of cheetahs were trapped and brought to the Mughal shikarkhana, and at one time there were more than a thousand of these swift cats in Akbar's household. A later chronicler, Muta'mid Khan, mentions that Akbar collected not less than 9,000 cheetahs during his 49-year reign. There were more than 200 keepers and trainers in charge of the *khasa* or imperial cheetahs. Slow breeders, and never in captivity, such a large scale capture of this never abundant animal must have heralded its extinction, though some number did survive up to 1950. Thereafter, the cheetah ceased to be in the Indian subcontinent.

Abu'l Fazl described Akbar's passion for the chase as a 'means of acquisition' of knowledge and as a pretext for making 'enquiries into the condition of the people and the army'. There are contemporary records both in print and in visuals to testify this Mughal emperor's great enthusiasm for hunting. Some visuals are preserved in the fragmentary *Akbarnama* manuscript housed in the Victoria and Albert Museum in London, painted by the leading painters of Akbar's painting atelier. A second *Akbarnama* manuscript completed in 1604, now divided between the British Library, London and the Chester Beatty Library, Dublin, has pictorial records of Abu'l Fazl's illuminating narrative. In addition, there are album paintings that show Akbar and Jahangir in hunting expeditions. Scenes of deer hunting invariably depict the cheetah. At least two early Mughal examples show a superb 'family portrait' of the cheetah and Akbar's personal involvement in cheetah trapping.

Abu'l Fazl has devoted two whole chapters in the *A'in-i-Akbari* to cheetahs where he has dealt with trapping, hunting, classification of cheetahs, quantity of food allocated to them, wages paid to their keepers and so on. The cheetahs were divided into eight categories and their ration of meat was graded accordingly—a first class cheetah got five seers of meat every day, and on the other side of the scale, an eighth class one got only two and three quarter seers. Hunting or animal slaughter was not permitted on Sunday and the animals were given double their daily portion of food on Saturday. The khasa, or the emperor's personal cheetahs, were given names according to their prowess and special characteristics. They held ranks and got splendid brocade coats and jewel studded collars. The animals were trained with great skill under the supervision of the courtiers. They were blindfolded and taken to the hunting field and made to sit on *goshaghani* carpets from Persia spread on horseback or small animal-driven carts or even in *doolie*s or litters. The captured cheetah was trained in 18 days. There was one instance when a cheetah without any training brought in the prey like a trained animal would have. According to Abu'l Fazl, Akbar himself took part in catching wild cheetahs in places like Hissar, Bhatinda, Fatehpur-Jhunjhunu, Nagaur, Jodhpur, Jaisalmer and Gwalior.

In November of 1560, on his way back to Agra from Punjab, Akbar decided to go from Sirhind to Hissar to do some cheetah trapping. Abu'l

Fazl writes: 'in a short time all the apparatus was got ready and sundry pits, which in Hindi are called *odi*, were dug according to a special plan (of Akbar's innovation). Several heads of *cita* were caught there before Akbar's cavalcade moved towards Delhi, though before this, many *cita* had been collected for the royal establishment, this was the first time that His Majesty hunted them in person'.

This incident was illustrated in the *Akbarnama*, in both the copies. The one in the Victoria and Albert Museum was designed by a senior master, Tulsi, and finished with colour by Narayan, while the one in Dublin was painted by Daulat. Akbar's new method of ensnaring the cheetah was by creating shallow pits, not more than two or three cubits, *gaz*, deep, fitted with trap doors which automatically closed as the cheetah fell in. The painting shows Akbar using this new method near Hissar. In part of the painting the emperor is shown putting a collar around the neck of the cheetah before shifting it to a cage. The one already caged is being carried by the attendants while huntsmen are shown preparing new pits in the lower foreground. The royal biographer has related another incident when Akbar collared a cheetah trapped in a hole in the neighbourhood of Gwalior in 1569, and brought it out. Standing instructions were that Akbar be informed as soon as a cheetah was trapped so that he could mount a swift horse and appear at the spot to catch it.

There is another painting on the same subject done in *nimqalam* (a technique in Mughal painting) which is partly coloured, by Aqa Riza, an emigre painter from Herat who worked as the director of Prince Salim's painting atelier during the last years of Akbar's reign. Though the royal figure is generally identified as Prince Salim, it in fact shows Akbar in his youth. This striking drawing is part of the collection of Prince Sadruddin Aga Khan.

This new method of trapping was so successful that on one occasion, no less than seven cheetahs were caught in one pit. It was, however, a case of love's labour lost—the first one to fall was a female 'on heat'. She fell accidentally into a pit and her six companions, 'unwilling to let her off, dropped in one after the other!' This is recorded in the *Akbarnama*. One of the six was Madan Kali who became Akbar's prized cheetah.

In another striking painting from the *Akbarnama* in the Victoria and Albert Museum painted by the great Mughal master Basawan and finished by Dharamdas, emperor Akbar's passion for hunting with cheetahs is graphic. The cheetahs were brought into the hunting ground in litters, on horseback, in cages hanging on either side of horses and camels and in bullock carts, perfectly matching Abu'l Fazl's description in the *A'in-i-Akbari*:

> For their conveyance, two litters (*mihaffa*) are hung over the back of the elephant, one litter on either side. On one litter one cheetah sits, looking out for a prey. Litters are also put on camels, horses and mules. Special carts made for the cheetahs are drawn by horses or cattle; or they are made to sit on horses; and sometimes are carried by men in *doolies*. The best cheetah His Majesty has, was Samand-Manik, he is carried on a *chaudol* and proceeds with much pomp. His servants fully equipped, run alongside; the *naqqara* is beaten in front, and sometimes he is carried by his men on horseback, the two ends of the pole of the *chaudol* resting on the necks of their horses.

In this painting, the leaping cheetah in front of Akbar's horse may well be Samand Manik, in his decorated *chaudol* (a gold litter), being carried by two retainers in the foreground. This wonderful painting by another leading painter, Khemkaran, illustrates the scene of Adham Khan's homage to Akbar near Sarangpuri in Malwa in 1560.

Akbar's intense and ongoing interest in cheetahs gave rise to cheetah lore and this cat began to appear everywhere—in court assemblies, paintings, book covers, textiles, carpets and metal ware. Individual cheetahs sitting majestically on bullock carts became a favourite subject of the painters. In the India Office Library in London, there are two seventeenth- and early eighteenth-century paintings of cheetahs seated on carts drawn by a pair of blackbuck, the very animal cheetahs were trained to hunt. Emperor Jahangir relates an episode of when a cheetah became friendly with a tame blackbuck in the royal household. Despite that, the cheetah continued to hunt blackbuck during expeditions.

Cheetahs came to be associated with musical melodies or the *ragamala*. A collared cheetah was often found seated in a temple in the pictorial representations of the *Bengali Ragini*. The French traveller Francois Bernier saw cheetahs and witnessed them hunting deer during his travels in India between 1656 and 1668. Jahangir once wanted to verify whether it was possible for an antelope that had escaped from the grasp of a cheetah to live if it had not been injured by the claws, and he discovered it was possible. When Akbar's two favourite cheetahs, Daulat Khan and Dilrang, drowned in the flooded Ganga near Patna in 1574, the royal biographer noted the incident in his book. That is how much the cheetah had permeated the Mughal court. Jahangir observed the capture of a female cheetah between the garden of Wafa and Nimlah, east of Jalalabad where 'the *zamindars* of that place, Laghmanis, Shalio and Afghans, came out and said that they did not remember nor had they heard from their fathers that a cheetah had been in that region for 120 years'. Had this animal begun to get rare even in the early seventeenth century?

Abu'l Fazl has related another incident about the extraordinary feat of a royal cheetah which once made a single 25-yard leap. This was at Sanganer, near Jaipur:

> One of the joyful occurrences was that at this time (July 1562) when the army was encamped at Sanganer, His Majesty according to custom, engaged in hunting. He was at this time much devoted to hunting with cheetahs and after assigning cheetahs to numerous parties he went off himself with some special attendants. It chanced that they loosed a special cheetah (*cita-i-khas*) called *Cita Najam* at a deer. Suddenly there appeared in front of them a ravine which was twenty five yards broad. The deer leapt into the air to the height of a spear and a half, and conveyed itself across. The cheetah in its eagerness took the same course, cleared the ravine and seized the deer. On beholding this astonishing occurrence the spectators raised a cry of amazement and there was great rejoicing and astonishment. The *khedive* raised the rank of that cheetah and made him chief of the cheetahs. He also ordered that as a special honour and as a pleasure to men, a drum should be beaten in front of the cheetah.

The *Akbarnama* in the Victoria and Albert Museum has a miniature that depicts this event, painted by Lal and finished with colour by Sanwalah.

No cub was ever born in Akbar's shikarkhana. Yet, on one rare occasion three cheetah cubs were born in Jahangir's menagerie, an occasion that got a special mention in the Mughal emperor's memoirs, the *Tuzuk-i-Jahangiri*. There is a splendid miniature in the Sadruddin Aga Khan collection, in brilliant colours done on fine linen that shows a pair of cheetahs with four young cubs playing in the wild. This is a rare example of a cheetah family shown in their natural habitat, very different from the familiar studies of tame cheetahs.

That is how the cheetah came into prominence and in many ways those years were the peak after which the cheetah population rapidly dwindled and finally, in the 1950s, this swift and wonderful wild cat became extinct in the Indian subcontinent.

THE HUNT AND THE WILDERNESS IN MUGHAL INDIA*

Mahesh Rangarajan

On 26 February in 1634 CE the Mughal ruler Shah Jahan chose to hunt at Palam—it is now the site of New Delhi's airport. The *Padshahnamah* records what happened in celebratory detail:

> The procession left Nurgadh and reached the hunting grounds of Palam, where descent was made to the imperial buildings that had been built in that area. For four days in that delightful spot (the *Padshah* or emperor) indulged in hunting and bagged much game. In one day, he himself shot forty black antelope with the royal gun called *Khasban*, and not one needed a second shot. This, of course, occasioned astonishment in all ... Taliba Kalim's poem recounted how, 'When Shah Jahan *Padshah*, lion hunter chanced to hunt in Palam/In one day with the gun *Khurasan* he shot 40 antelope without a single shot missed.'

Using the musket was second nature to the Mughals by the seventeenth century, but in the time it took to fire a single shot, it was said, a good archer could let loose as many as six arrows. This common maxim did not stop successive rulers from using a range of weapons other than the bow and arrow.

As in previous ages, the hunt was more than a means of pleasure or relaxation. Advances in weapon technology over the centuries had not eroded the symbolic importance of killing wild, free-ranging animals. For a militarised, horse-borne nobility, it was a means to tone the body, to test one's nerves and learn the fine art of stalking. A vassal

* Mahesh Rangarajan, 'The Hunt and the Wilderness in Mughal India', in Mahesh Rangarajan, *India's Wildlife History: An Introduction*, New Delhi: Permanent Black, 2005, pp. 11–21.

of the Mughals from the hills of Kumaon, Rudra Deva, felt that royal operations against wild prey were 'the same as slaying enemies': killing ferocious beasts or the deer that raided corn were acts of religious merit.

Certainly, hunts were used to camouflage major armed expeditions, especially against recalcitrant vassals. Though established first in India by Babur, a disinherited central Asian prince, the Mughal Empire was really set on a firm footing by his grandson Akbar who ruled for nearly half a century, ending in 1605. His biographer, Abu'l Fazl, claimed that the hunt was a means of gathering intelligence about the state of the realm. At a different level, the Mughals developed hunting into a ritualised activity, laden with political meaning. Arrangements were overseen by a Hunt Master or *Mir Shikaran*. The Padshah's party often hunted with a huge retinue of soldiers, who drove the game animals into a central staging ground. The half-moon formation or *qamargah* formed by the troops required a deep knowledge of the terrain and movements of the quarry. Once encirclement was complete, the emperor chose the noblemen who would personally accompany him in the actual act of hunting. At times, despite all the arrangements, the hunters moved on foot, exposing themselves to great personal danger, especially when pursuing a lion or tiger. When Jahangir was saved from near certain death in a lion hunt at Bari by a Rajput prince, he attained the stature of legendary Indian and Persian kings of yore. In this case, it also further reinforced, through the Rajput's personal bravery, the Mughal–Rajput alliance that was the pillar of the empire.[1]

For anyone trying to piece together patterns from the past about the changing equations between humans and wild animals, the Mughal Empire provides a treasure trove of material. There are memoirs with names, dates and places and crucially, paintings and sketches to go with particular events. Unlike most writers in Sanskrit who were vegetarian brahmins, the Mughals as keen meat-eaters and hunters, were particular about observing what they hunted. Babur, the first of the Mughal rulers, died in 1530, but his memoirs, written in Turki and later embellished

[1] See Milo Beach, Ebba Koch and W. Thackston, *King of the World: The Padshah Namah*, New York: Arthur M. Sackler Library, 1997, pp. 6–7; M. M. Hara Prasad Shastri, trans. and ed., *The Syainikasastra of Rudra Deva of Kumaon*, New Delhi: Reprint Publishers, 1982, p. 83; A. Roger and H. Beveridge, trans. and ed., *The Tuzuk i Jahangiri*, London: Royal Asiatic Society, 1909, pp. 185–88.

with portraits, recorded his sense of awe at the birds and beasts of the new country he had come into. His grandson Akbar was a short man with a pockmarked face, in all probability dyslexic. His lack of literacy did not hinder his ability to forge together an empire, which, by the time of his death in 1605, stretched across half of South Asia. More importantly, his gifted hagiographer, Abu'l Fazl Allami, with his extraordinary gifts of record-keeping, wrote in Persian about the events in the emperor's life.

There were notes of continuity with the past: Rudra Deva's earlier Sanskrit work on falconry and the Persian writings of the emperors show remarkable similarities. This is even clearer in the case of arguably the best manual of the hunt in Sanskrit, the *Manasollasa* of the Chalukya rulers in the Deccan plateau, composed in the twelfth century. The text details certain techniques of deer and antelope hunting, including coursing with cheetahs or trapping a wild male blackbuck by using a tame decoy animal. There is an uncanny similarity between such tactics and those known to Akbar. Court artists portrayed many hunts that were recounted in great detail in memoirs. The rulers took a huge share of surplus farm produce and had enormous resources at their command and in turn, patronised artists of repute. *Ustad* Mansur, the first man to accurately portray the magnificent Siberian crane, was only the most notable in a large team of artists. Their realistic pictures were not only rich in colour, but also convey a sense of the drama in everyday life.[2]

Imperial bags could be prodigious. Despite the huge bags, there was an abundance of wildlife. In the first 12 years of his reign alone, Jahangir killed over 17,000 animals. These included as many as 889 *nilgai*, 86 tigers and lions and 1,670 gazelle and antelope. The number of large cats slain works out to just about seven a year, hardly a large figure when compared with more recent instances of slaughter. Wild meats were highly prized, perhaps explaining the huge numbers of birds and herbivores hunted down. The Mughal palate was used to 35 to 40 meat dishes at a proper meal, including venison, quail and partridge. The hunt was not only a means to put fresh meat on the plate. It also symbolised

[2] Divyabhanusinh, *The End of a Trail: The Cheetah in India*, New Delhi; Banyan Books, 1995, pp. 35–77. This chapter is perhaps the best work yet on natural history in the Mughal period; Ebba Koch, *Dara Shukoh Shooting Nilgais: Hunt and Landscape in Mughal Painting*, Smithsonian Occasional Paper No. I, Washington, 1997.

the ability of the ruler to overcome 'noxious animals'—a term used by Jahangir for a huge tiger he killed near the fort of Mandu. The success or failure of a hunt on the eve of a major military expedition was taken as an omen of what lay ahead. Descriptions of habitat at the time bear little resemblance to the areas concerned in their present state, but the Mughal accounts give us portraits of past ecologies.

Babur wrote of 'great masses' of rhinos between the Indus river and the city of Bhira. It was difficult to pierce the rhino's hide with arrows, but there was a flourishing trade in rhino horn cups, which were said to be useful in detecting poison. The horn does not appear to be widely used as an aphrodisiac at this time. Elephants were caught later in the sixteenth century from parts of central India like Hoshangabad, Raisen and Chanderi. At one stage, there were over 12,000 elephants in the possession of emperor Jahangir (d. 1628) and his nobles or *mansabdars*. In north Bengal, a traveller through the great grasslands along the rivers saw huge herds of wild buffalo that, 'never stirs from his position in fear of the musket, arrow or sword.' While such species still survive in parts of their range to this day, the same cannot be said of that most swift and elegant of the big cats, the cheetah. Unlike the *chital, nilgai* or *sambhar,* for which the Mughals used Hindavi names, they had a Persian word for the cheetah: the *yuz*. Akbar kept over a thousand at one time, using them to hunt the antelope, the gazelle and even deer. His son was to breed them in captivity, a rare event that he dutifully recorded in detail. But in the Mughal imagination, it was the tiger and the lion that played a key role. The latter figured in the flag of the empire, and was common enough across much of the scrub forests and dry savannahs of north India. There are no recorded instances, in either the paintings or memoirs, of either lions or cheetahs being brought from Africa. Tigers were to be encountered even in the ravines along the river Yamuna close to the premier cities of Agra and Delhi.[3]

Mughal techniques of observation and record could be the envy of a modern zoologist and the emotions expressed would be familiar to

[3] On Jahangir, see *Tuzuk*, pp. 362–63, 369; A. S. Beveridge, trans. and ed., *The Babur Namah*, New Delhi: Oriental Reprint, 1970, pp. 489–90; on elephants, Irfan Habib, *An Atlas of the Mughal Empire*, Cambridge: Cambridge University Press, 1982, p. 38; on Bengal, J. Sarkar, 'A Description of Bengal in AD 1609', *Bengal Past and Present* 35: 69–70, January–June 1928, pp. 143–46.

today's animal lover. Jahangir described an American turkey as 'larger than a pea hen and smaller than a peacock.' An African elephant was labelled an elephant from across the seas or *dariyayi* hathi and clearly differentiated from its Indian counterpart. Individual animals at the court were known and their deeds celebrated. Akbar inspected the huge tusker called Gajraj; another ruler built a tower in memory of his favourite blackbuck. Ibrahim Adil Shah, the ruler of Bijapur, took this attachment to individual creatures to new heights. Compelled to cede the cow elephant, Chanchal, to the Mughals as tribute, he thought of the effect this might have on her mate, the tusker Atash Khan, his favourite mount. He even imposed a moving poem on the pangs of separation felt by elephants estranged in love![4]

The Mughals presided over an empire of great diversity, both ecological and cultural. But the extension of agriculture was at the heart of the ability of the empire to survive and thrive. Despite war booties and tribute, the taking of the surplus grain grown by the cultivator, or the *ria'yya* was what underwrote the splendours of court life. Peasants hacking down the jungle to make way for farm land were rewarded by their revenues being written off for the first few years. In practice, the rulers drew on a range of skills and forms of wealth other than cash revenues. The *peshkash* or gifts given as tribute provide an insight into the wealth of the wildlife. Elephants were received as tribute from some north Indian villages in lieu of cash; from the Mishmi hills in the north-east came musk from the musk deer. Aurangzeb knew of five sites in Assam where, 'if desired 500 or 600 elephants could be captured annually.'[5]

What is significant is that the total population of India at the time of Akbar's death was estimated to be not less than 150 million and at most, 170 million. If the former figure is correct, it means the rates of the growth of human numbers remained low until well into the nineteenth century. In 1850, the population of India was around 250 million—a rise of around 100 million at most in 25 years. There were vast pastures for cattle, firewood was abundant in most parts of the realm, and many areas had higher rainfall levels than at present. It is quite possible that more land was forested than was once thought to

[4] Divyabhanusinh, *The End of a Trail*, pp. 40–42, 64–65; M. Zebrowski, *Deccani Paintings*, New Delhi: Roli Books International, 1983, pp. 96–98.

[5] Habib, *Atlas*, Sheets 13B and 15B.

be the case. Even in the core of the empire in the revenue district of Agra only one out of four acres was under the plough. People were even fewer in areas of high rainfall, in dense forest or in places plagued by crop-raiding elephants. Forests were undoubtedly being cleared, but in large regions of central India and the Chhotanagpur plateau, uncultivated lands and forest covered a greater expanse than fields and farms. Land was more abundant than labour, the latter being essential to clear and keep back the jungle. Landed elites and princes had to tread with caution: peasants could simply flee tax collectors and move off into the forest. Specific tracts were, of course, being denuded of tree cover and bushes. Of Dohad and Rajpipla in Gujarat, a chronicler would write by the 1760s that though wild elephants were once caught there, the passes had since been closed by human settlements. This indicates that valleys, low-lying areas and the plains were the first to be denuded for agriculture, fragmenting the home ranges of large wild animals. Gujarat with its low rainfall was easier to denude of tree cover than other more moist provinces.[6]

The forest and savannahs were many things at the same time: sites of hunts, great and small, potential revenue-yielding arable land or simply an impediment to military operations. The jungle was finally also a place to commune with the supernatural, a place where dark spirits were harnessed with the powers of meditation. The wild spaces could be a hunting ground where prior imperial rights were asserted and enforced. Certain lands had to be set aside to ensure the success of royal hunts. Unlike the elephant forests of the Mauryas, the Mughal locations are easy to map and it is clear there were a range of imperial hunting grounds in different *suba*s or provinces. Rupbas in Rajasthan was one such place, virtually uninhabited, a place where the imperial cheetahs could course the antelope 'without molestation'. An English traveller and merchant, Edward Terry, found property in game exercised very lightly. 'The whole kingdom as it were a *forrest*', he wrote, 'for a man can travel in no direction but see them, and except it be within a small distance of the king, they be every man's game.' Terry was

[6] Tapan Raychaudhuri and Irfan Habib, *The Cambridge Economic History of India, c. 1200–1750*, Vol. I, Cambridge and New Delhi: Cambridge University Press, 1982, pp. 163–70; K. K. Trivedi, 'Estimating Forests, Wastes and Fields c. 1600', *Studies in History* 14: 2, August 1998, pp. 301–11; Habib, *Atlas*, Sheet 7B.

probably comparing the Mughal hunting places to the deer forests in his native England where property rights were already enforced much more sternly. In India, the concept of property in jungle and woodland was not so absolute. For instance, at the *shikargah* or hunting ground near Agra, local residents could trap quail, hare and partridge, though not larger wild quarry. Access was open, but the kind of harvest of wild meats or game taken was regulated. Peasants were pressed into service to corner a pride of lions for Shah Jahan in the Deccan. Military forces often brought sawyers, woodcutters and road builders in their train, working hard to deny rebels and runaway peasants cover and to facilitate the large-scale advance of imperial troops. In times of war, denudation was a routine affair.

Conversely, the scrub was often a re-growth of secondary vegetation in abandoned fields.[7] Sufi *kalandars* or heterodox Muslim mystics camped in the forest to meditate. Paintings from Mughal India and from the polities of the Deccan depict holy men, often of various Hindu sects, seated peacefully next to lions or tigers. Muslim mysticism took the older track of Hindu religious figures that had retreated into the forest in search of the self. Akbar's *jazbah* or religious frenzy when out on a hunt for wild asses in Sindh province reinforced Mughal claims of divine status for the emperor. Persian pre-Islamic beliefs in the ruler as 'a mountain of light' merged with Indian customs like sun worship. After a second such trance-like experience in 1578, the emperor suspended the shikar that was in progress on the banks of the river Jhelum. He even forbade hunting on certain days. Earlier, an equally remarkable Kashmiri Sultan, Zain-ul-Abidin gave up meat-eating and tried to dissuade his nobles from hunting. Such efforts could hardly be expected to succeed. But what they probably drew on was a revulsion against slaughter that was widespread and took a religious form. Under Mughal rule, no killing was permitted at sacred sites of the Jains like the Shetrunjaya hill, Saurashtra or Parashnath in Chhotanagpur. The mostly mercantile community of the Jains, a sect already over two

[7] W. Foster, ed., *Travels in India, 1583–1619*, London: Hakluyt Society, 1927, pp. 296–97; F. Bernier, *Travels in the Mogol Empire, 1656–68*, 2nd revised edition, trans. and ed. A. Constable, New Delhi: Oriental Reprint, 1983, pp. 374–76; Beach et al., *King of the World*, p. 110; Zebrowski, *Deccani Paintings*, pp. 111–12.

millennia old, also ran hospitals for animals and birds in the towns and cities. Local customs and practices were clearly given their due place.[8]

The picture of Mughal India in terms of the human impact on nature emerges more clearly if contrasted with experiences of contemporary Europe. The dodo was brought to Jahangir's court via Portuguese-ruled Goa: Mansur's was the first accurate record of the bird. By the end of the seventeenth century, the species had disappeared forever, both from being hunted and from the introduction of new animals by French settlers. Within Europe, the levels of human pressure took their toll, with the great wild cattle, the auroch, becoming extinct in central Europe by 1627, the last year of Jahangir's rule. The destruction of the island-specific dodo was easy given its relative isolation from newer pressures in the game of survival. But the auroch's case throws light on the ability of the European aristocracy to wipe out an animal in its entirety. In India, there still existed much larger stretches of contiguous habitat even for the larger land mammals. Further, unlike in contemporary Europe, there were as yet no rewards for killing flesh-eating beasts that sometimes menaced cattle or people. Such pressures were unknown in Mughal India.[9]

The Mughals took aboard older traditions of the hunt but added to it significant new features. Their penchant for record-keeping in both the written word and in the form of portraiture arose from a heightened sense of curiosity and a deep sense of history. The picture that emerges is of vast areas of woody, scrub jungle lightly populated by people and cattle, abounding in antelope, gazelle and lions. Denser forest was known in many areas but whether there were any primeval forests untouched by human hand is doubtful. Fire, axe and plough were reshaping vegetation patterns even as species like the elephant and the cheetah were being trapped on a larger scale than before. The ebb and tide of forests and the newly cultivated spaces were themselves a prominent feature of the times. Habitat of large mammals like the elephant and the rhino was broken up—a harbinger of larger changes in the land. Horsemanship, constant military campaigns and wars brought

[8] Koch, *Dara Shukoh*, pp. 11–12; R. Fitch in Foster, *Travels*, p. 25.

[9] R. Silverberg, *The Dodo, The Auk and the Oryx*, London: Puffin, 1973, pp. 29–37, 54; M. Alvi and A. Rahman, *Jahangir, The Naturalist*, New Delhi: National Institute of Sciences, 1968.

the ruling elites in close contact with the outdoors. What was striking was the range of observations: of the behaviour of falcons, the hunting techniques of the tribals and of the myriad ways of tracking down tigers. The hunting grounds of the emperor were not absolutely out of bounds for subjects, though there were selective restrictions. Commentaries routinely mention tribal or local techniques without denigrating or criticising them. Tactics like the capture of cheetahs in pit-fall traps or the trapping of francolins probably evolved in association with local catchers. Despite the disintegration of the empire after 1707, several of its symbols lived on, if with new meanings. Falconry, cheetah coursing, horsemanship and archery remained the main pastime of landed gentry across much of north and central India for much longer. Rulers across India commissioned paintings and portraits celebrating victory over the beasts of the forest. Mughal power had waned; its legacy lived on.

NOTES ON THE CONTRIBUTORS

Salim A. Ali was an Indian ornithologist and naturalist, also known as the 'birdman of India'.

M. A. Alvi was Professor with National Institute of Sciences of India, New Delhi.

Meena Bhargava is Associate Professor at the Department of History, Indraprastha College for Women, University of Delhi, New Delhi.

Bhangya Bhukya is Associate Professor at the Department of History, School of Social Sciences, University of Hyderabad, Gachibowli, Hyderabad.

Asok Kumar Das is a senior art historian and specialist in Mughal art. He retired as Tagore National Professor, Indian Museum, Kolkata. Currently, he is Academic Advisor to the Nehru Trust for the Indian Collections at the Victoria and Albert Museum, Kolkata.

Simon Digby was a Fellow at Wolfson College, Oxford and a recipient of Burton Medal of the Royal Asiatic Society.

Mayank Kumar is Associate Professor at the Department of History, Satyawati College (Evening), University of Delhi, New Delhi.

Archana Prasad is Professor with Centre for Informal Sector and Labour Studies, School of Social Sciences, Jawaharlal Nehru University, New Delhi.

A. Rahman was Professor and Founder Director of National Institute of Sciences of India, New Delhi from which he retired in 1985.

Mahesh Rangarajan is Professor of History and Environmental Studies at Ashoka University, Sonepat, Haryana.

Chetan Singh, formerly Director, Indian Institute of Advanced Study, Shimla, is Professor of History at Himachal Pradesh University, Shimla.

Tripta Wahi retired as Associate Professor from the Department of History, Hindu College, University of Delhi, New Delhi.